Delta – The Structure of Change

Sandeep Chavan

Published by Gyrus Vision, 2026.

DELTA – THE STRUCTURE OF CHANGE

First edition. March 2, 2026.

ISBN: 979-8232686536

Written by Sandeep Chavan.

Table of Contents

To those who choose to observebefore they conclude— and to those willing to replace beliefwith clarity.

Imbalance is not error.It is information.

Resolution is not judgment.It is structure restoring itself.

Opening Note

This book did not begin as an idea. It began as a discomfort.

For years, I studied, taught, observed, and explained the world using the language that physics, philosophy, and common sense had handed to us—force, motion, cause, effect, observer, control. It worked well enough to solve problems, pass exams, build machines, and even predict outcomes. And yet, something fundamental always felt missing.

Reality behaved more intelligently than our explanations.

A soap bubble demanded skill, not force. A fluid responded to gentleness more than pressure. Friction did not oppose motion—it *regulated* it. Chaos did not appear randomly—it arrived when something exceeded its capacity to resolve. Coherence was never accidental—it emerged only when conditions were just right.

Slowly, it became impossible to ignore a simple truth:

Nothing moves because it is pushed. Things move because something is misaligned.

That misalignment—what I call delta (Δ)—turned out to be everywhere.

Not as a number. Not as a disturbance. But as a *condition*.

Energy can exist endlessly without doing anything. Entire universes could remain silent if nothing were misaligned. Motion, change, flow, life, thought—none of these arise from energy alone. They arise only when delta begins to resolve.

Once seen, this was impossible to unsee.

Delta explained why:

- large imbalances create chaos,

- tiny imbalances create jumps and noise,
- and only the *right* imbalance creates coherence.

It explained why time feels different in different systems. Why equilibrium is dynamic, not static. Why the observer does not create reality, yet still experiences it meaningfully. Why nature behaves perfectly well without us—and why we still matter.

This book is not an attack on existing science. It is not a rejection of equations, laws, or discoveries.

It is an attempt to place them in the correct order.

Force becomes a consequence, not a cause. Motion becomes resolution, not action. Stability becomes alignment, not rigidity. Loss becomes redistribution, not destruction.

And the observer—often made divine, mysterious, or central—is returned to his rightful place: a well-organized, sustained, aware participant in an ongoing dynamical world.

What follows is not speculation. It is a re-reading of reality that starts from what is most honest: experience that refuses to lie.

This book asks only one thing of you:

Do not try to believe it. Do not try to reject it.

Just observe the world—carefully, patiently—and ask yourself whether delta is not already doing everything you see.

If it is, then this book will feel less like learning something new and more like finally naming what was always there.

Preface

Most confusion about reality does not come from a lack of information. It comes from the wrong framing of change.

We are taught to think in terms of objects, forces, intentions, and causes. We search for actors behind events and meanings behind outcomes. When results surprise us, we label reality complex, mysterious, or unfair. The assumption beneath these reactions is simple: something must be pushing something else.

This book takes a different approach.

Delta is not concerned with what things are, but with what shifts. It focuses on imbalance, buildup, threshold, and resolution—the quiet structural movements that determine outcomes long before they are named, judged, or interpreted. Instead of asking what controls reality, it asks how structure reorganizes when it can no longer sustain tension.

There is no divine observer in this framework. No privileged reference frame. No metaphysical refuge when explanations feel uncomfortable. Only structure, resolution, and experience—each emerging from the same underlying logic.

This book was written to remove mystery where none is required—and to replace force-based thinking with resolution-based understanding. It does not compete with science or philosophy. It clarifies the assumptions they often leave implicit.

If this perspective feels unfamiliar, that is expected. Delta does not add complexity; it removes excess interpretation. What remains may feel stark at first—but it is also precise.

You do not need agreement to benefit from this lens. You need only careful observation.

Once you begin to observe differently, something subtle becomes clear:

Reality is not erratic.

It is structural.

This book is an invitation to see that structure without distortion.

How to Read This Book

This book is not meant to be believed. It is meant to be used.

Delta is not a theory to memorize or a philosophy to defend. It is a way of seeing structure—how change forms, how imbalance accumulates, and how outcomes resolve long before they are interpreted. It invites you to observe differently, not to adopt a new doctrine.

If you approach this book expecting formulas, predictions, or metaphysical revelations, you may feel disoriented at first. That response is natural. The purpose here is not to confirm familiar habits of thought, but to gently retrain perception.

Read slowly. The chapters follow a progression, but understanding does not always move in straight lines. Some ideas will feel immediately clear; others may take time to settle. Do not force agreement. Let repetition and contrast do their work. Often, a later section will illuminate something earlier. That is not accidental—it reflects how resolution itself unfolds.

Do not look for a final answer hidden within these pages. This book does not reveal a secret behind reality. It removes unnecessary mystery from processes that already operate structurally. Meaning appears after resolution, not before it.

Use the everyday examples—traffic, fatigue, learning, conflict—as anchors. They are not metaphors. They are confirmations. Structure behaves consistently across scales.

You may notice that your intuition shifts before your conclusions do. Situations begin to appear less as battles of force and more as patterns of buildup and release. That subtle change matters.

Read with suspension, not faith. Observe carefully. If something unsettles you, hold it. If something feels obvious, ask why it was overlooked.

Reality was never chaotic.

It was simply unread.

Core Thesis

Delta (Δ) is the fundamental driver of all dynamism. Energy by itself does nothing. Existence does not unfold merely because energy is present. *Manifestation begins only when delta starts resolving.*

Delta is not a quantity to be measured first, nor a disturbance added later. It is a *condition of misalignment*—between regions, states, configurations, or possibilities. Until such misalignment exists, energy remains dormant. When delta appears, nature does not hesitate, decide, or wait for instruction. Resolution begins immediately, and with it arise motion, change, structure, and time.

What we commonly describe as force, motion, pressure, flow, friction, stability, or chaos are not primary causes. They are expressions of how delta is introduced, sustained, and resolved within a system. When resolution proceeds smoothly, coherence appears. When resolution is overwhelmed, chaos follows. When resolution is starved, reality fragments into discontinuous jumps and fluctuations.

This book argues that *resolution—not force—is the true engine of becoming.*

From this standpoint, several familiar concepts must be reordered, not rejected:

Force is not fundamental; it is a macroscopic appearance of unresolved delta. What we perceive as pushing, pulling, attraction, or resistance emerges when alignment cannot be maintained without redistribution.

Motion is not an externally imposed action; it is delta resolving through available pathways. Objects do not move because they are acted upon; they move because misalignment seeks release.

Equilibrium is not stillness or rest. True equilibrium is aligned resolution—a state in which delta continues to resolve without accumulation or rupture. Stability is dynamic, not static.

Chaos is not randomness or disorder by default. Chaos arises when delta exceeds a system's capacity to resolve coherently, fragmenting resolution across scales. Disorder is not the opposite of order; it is failed coherence.

The observer occupies no privileged or creative position in this process. Nature evolves independently through delta and resolution. The observer is an organized participant, not a creator—an internally coherent system that experiences manifestation when its own alignment allows access to ongoing resolution. Meaning, interpretation, language, and manipulation arise *after* alignment, not before.

This framework applies without exception across scale and domain: from quantum transitions to fluid flow, from friction to life, from cognition to social dynamics. No new forces are introduced. No metaphysical assumptions are required. The same principle governs all: *delta appears, resolution follows, form emerges.*

The purpose of this book is not to replace existing laws, but to restore their causal order. When delta is placed first, everything else—equations, models, intuitions, and experiences—falls into coherence.

Once delta is seen clearly, reality stops appearing mysterious. It becomes intelligible, continuous, and exact—doing precisely what it must, and nothing more.

Part I — Why Delta Was Missed

This opening part of the book does not introduce new ideas immediately. Instead, it creates the necessary intellectual space to let old assumptions loosen their grip. For centuries, our understanding of the physical world has progressed by describing *what happens* with extraordinary accuracy, while quietly bypassing the question of *why it happens at all*. The result is a worldview that is mathematically powerful, technologically successful, and yet experientially unsatisfying.

This part begins by examining this inversion.

Modern physics typically starts with force, interaction, or laws of motion, treating imbalance as a secondary detail. Equations are presented as causes, even though they are fundamentally records of outcomes. This ordering was not inevitable; it emerged from historical necessity, mathematical convenience, and early experimental success. Over time, the language of force hardened into ontology, and imbalance was reduced to a parameter rather than recognized as the origin of dynamism itself.

As a result, a quiet discomfort appears whenever theory meets lived reality. Phenomena work, predictions hold, yet intuition resists full agreement. Motion feels imposed rather than emergent. Stability feels artificial. Chaos appears accidental. This discomfort is not a failure of intelligence—it is a signal that causality has been placed upside down.

The second chapter strips the problem to its simplest form: energy can exist indefinitely without producing motion, change, or structure. Static systems feel inert not because energy is absent, but because nothing is misaligned. Potential alone does not ignite reality. Something else must invite change. That missing invitation is delta.

The third chapter introduces delta carefully, not as a number to be calculated, but as a condition to be recognized. Delta is misalignment, not difference; tension without direction. It precedes force, motion, and even time. Without resolution, delta is experienced as strain. Without delta, resolution collapses into silence. Dynamism exists only where both are present.

By the end of this part, you are not asked to accept a new theory. They are simply prepared to see the world differently—by reversing the causal order that has quietly shaped our explanations. Unlearning here is not destruction; it is alignment.

Chapter 1 — The World Explained Backwards

When Understanding Feels Technically Right but Intuitively Wrong

There is a quiet moment that most people who have studied science deeply recognize, even if they rarely speak about it. Everything appears correct. The equations balance. Predictions match experiments. Technology works. And yet, somewhere beneath this technical confidence, a subtle discomfort persists.

The world behaves more intelligently than our explanations allow.

We are told that motion begins because a force acts, that stability exists because forces balance, and that change occurs when forces overcome resistance. These statements are not false. They are useful, powerful, and operational. They allow us to calculate trajectories, build machines, and send probes across the solar system. But usefulness is not the same as causality.

At some point, understanding starts to feel backward.

A soap bubble does not respond well to force, but to gentleness. A fluid does not reveal its nature under pressure, but under flow. Friction does not simply oppose motion; it regulates it. Chaos does not appear randomly; it emerges when something exceeds its capacity to remain aligned. None of this feels accidental, yet force-based explanations struggle to express *why* these behaviors feel so natural when experienced directly.

The discomfort grows precisely because nothing is technically wrong. The models work. The results repeat. The predictions hold. But the explanations feel like they begin after the real story has already started.

Equations tell us what happens once motion exists. They describe relationships between quantities once change is underway. They are excellent mirrors of behavior—but mirrors do not initiate movement. Somewhere before force, before motion, before calculation, something must already be unsettled.

Most learners suppress this discomfort. They assume it reflects a personal limitation rather than a structural one. Others intellectualize it away, accepting that "this is just how physics works." A few mistake the gap for mystery and push it into metaphysics or philosophy. But the discomfort itself is neither ignorance nor mysticism. It is a signal.

It suggests that we may have learned to explain the world from the wrong end.

This chapter does not ask you to doubt science. It asks you to notice where explanation stops short of experience. Not to discard equations, but to question their starting point. Not to reject force, but to ask what must exist *before* force becomes meaningful.

Unlearning here is not destruction. It is reorientation.

Before motion can be calculated, something must already be misaligned.

1.1 — How Force Became the Starting Point

To understand why modern physics begins with force, we must first acknowledge that this choice was not foolish—it was practical. Early science did not set out to explain reality in its deepest sense; it set out to measure, predict, and control. Force offered a tangible handle. It could be felt, inferred from motion, represented geometrically, and later expressed mathematically. In a world hungry for reliability, force was an accessible entry point.

Force made nature legible.

When objects moved, something seemed to act on them. When structures held, something appeared to balance them. When motion changed, something must have caused that change. These intuitions were not wrong; they were incomplete. But because they worked so well, they hardened into first principles. What was initially a convenient starting point quietly became the assumed origin of all dynamics.

Imbalance, meanwhile, was treated as secondary. It appeared only as a condition applied *to* force—unequal forces, net force, unbalanced forces—never as the root cause itself. In this framing, imbalance was something forces responded to, rather than something that preceded and demanded resolution.

This ordering mattered more than anyone realized.

Once force was placed at the foundation, explanations naturally flowed outward from it. Motion was said to begin when force acted. Stability existed when forces canceled. Change occurred when forces exceeded resistance. These statements were operationally true, but they subtly inverted causality. They described how systems respond once something is already wrong, without asking what made the system wrong in the first place.

The deeper issue is not that force was used, but that it was elevated from descriptor to origin.

As physics advanced, the language of force persisted even when its conceptual footing weakened. Fields replaced contact forces. Potentials replaced direct pushes. Constraints, gradients, and variational principles entered quietly, doing much of the real work—yet force remained the headline explanation. The vocabulary stayed familiar even as the machinery underneath grew more abstract.

Over time, the habit solidified: we learned to start explanations where calculation became possible, not where dynamism actually began.

This is why force-based thinking feels so effective and yet subtly unsatisfying. It tells us how motion evolves once it exists, but not why motion must exist at all. It explains regulation, not ignition. It assumes a disturbance without naming its origin.

Force did not become fundamental because it was deepest. It became fundamental because it was first to be measured.

That historical convenience shaped an entire worldview—one that works remarkably well, yet leaves a quiet question unanswered:

What had to be misaligned before force ever appeared?

This question does not negate force. It simply asks us to place it after, not before, the true beginning of change.

1.2 — Equations Describe Outcomes, Not Causes

Equations are among humanity's most powerful achievements. They compress vast patterns of behavior into precise relationships. They allow prediction without repetition, foresight without trial, and control without direct intervention. When an equation works, it inspires confidence—and rightly so. But success has a subtle side effect: it tempts us to confuse description with causation.

An equation never makes anything happen.

It records what happens *once something has already begun*. It maps relationships between quantities that are already in motion, already interacting, already changing. Even the most fundamental equations in physics presuppose a world in which dynamics are underway. They begin after the first disturbance, not at its origin.

Take any familiar law. It relates force to motion, pressure to flow, energy to configuration. But none of these laws explain why force should arise at a particular moment, why pressure should differ here and not there, or why energy should reorganize instead of remain dormant. The equation enters the scene after reality has already leaned away from balance.

This distinction is easy to miss because equations feel authoritative. They close the loop. They give finality. Once written, they appear to answer the question fully. But what they actually do is summarize resolution, not initiate it.

This is why so many physical laws feel timeless and silent at the same time. They do not care about beginnings. They do not ask why. They assume change and tell us how it unfolds. In doing so, they perform their role perfectly—but only within a limited scope.

The confusion arises when this scope is forgotten.

Over time, mathematical relationships began to stand in for explanations. If an equation accounted for behavior accurately, it was treated as the cause of that behavior. The logic quietly reversed: instead of saying "this equation describes what happens," we began to think "this equation explains why it happens." The difference is subtle, but foundational.

Consider equilibrium conditions. An equation can show that forces balance, pressures equalize, or energies minimize. But equilibrium itself is not caused by the equation. It is the result of a system having nothing left unresolved. The equation merely captures the end state of that process.

The same is true for motion. Equations describe trajectories, velocities, accelerations—but they do not explain why motion begins rather than not. They do not tell us why a system leaves rest, why it prefers one path over another, or why it reorganizes instead of remaining still. Those questions are answered before mathematics enters.

When explanation begins with equations, causality is quietly shifted downstream. The origin of change is assumed, not examined. As long as predictions succeed, this omission remains hidden. But when intuition resists—when lived experience feels richer, more conditional, more sensitive to rate, context, and gentleness—the gap becomes visible.

Equations are not wrong. They are incomplete by design.

They were never meant to tell us why reality becomes dynamic. They were built to tell us how dynamics behave once they exist. Treating them as causal engines asks them to do a job they were never created to perform.

This chapter does not argue against mathematics. It restores it to its proper place: as a mirror of resolved behavior, not the source of resolution itself.

Before equations can speak, something must already be unsettled.

1.3 — The Historical Accident of Force-Centric Thinking

Force did not become central to our understanding of the world because it was the deepest truth. It became central because it was the first truth we could reliably handle.

Early science emerged in an environment where survival, construction, navigation, and warfare demanded immediate, repeatable results. The questions that mattered were practical: *How far will it go? How fast will it fall? How much weight will it hold?* Force answered these questions directly. It could be inferred from motion, measured through effect, and later expressed with mathematical clarity. It fit the tools of the time and the needs of the moment.

This success was decisive.

Once force-based descriptions proved effective, they shaped not only methods, but mindset. A framework that predicts well quickly earns trust, and trust hardens into ontology. Over time, what began as a convenient modeling choice quietly transformed into an assumption about how reality itself must be structured.

The deeper causes were never denied; they were simply postponed.

As scientific thinking evolved, the machinery underneath force-based language grew increasingly sophisticated. Potentials, gradients, fields, variational principles, and boundary conditions began doing most of the conceptual work. In many advanced formulations, force all but disappeared as a fundamental entity. Yet the language remained. We continued to *speak* in terms of force even when we no longer *thought* in terms of force.

This mismatch created a strange situation: the mathematics moved forward, but the explanatory narrative stayed behind.

What had once been a helpful simplification became an inherited habit. New generations learned to start explanations where calculations were easiest, not where dynamism originated. Imbalance was absorbed into symbols. Gradients were treated as mathematical conveniences rather than physical conditions. Resolution was encoded but never named.

The accident was not ignorance—it was timing.

Physics matured before we had the conceptual tools to distinguish what triggers change from what governs change once it begins. Force was asked to perform both roles. It did the second brilliantly and the first poorly, but its success masked the distinction.

Over centuries, this ordering became self-reinforcing. Textbooks, pedagogy, and intuition all aligned around force as the beginning. Any discomfort that arose was attributed to complexity, abstraction, or human limitation—not to a misplaced starting point.

And yet, the signs were always present.

Phenomena where force mattered less than rate. Systems where gentleness outperformed strength. Processes where stability depended on timing rather than magnitude. Situations where equations worked but explanations felt hollow.

These were not anomalies. They were hints.

Force-centric thinking did not fail; it stopped one step short. It captured how nature regulates motion, but not why nature becomes dynamic at all. That missing step—imbalance as a primary condition—was never formally rejected. It was simply left unnamed.

This chapter does not accuse history of error. It recognizes circumstance. A framework born for control and prediction cannot be blamed for overlooking ignition. But now, with hindsight and accumulated insight, the omission becomes visible.

What once served us well has quietly limited how deeply we explain.

The world was not explained incorrectly. It was explained from the middle, not from the beginning.

1.4 — Why Discomfort Appears When Theory Meets Reality

The most revealing moments in understanding do not arise when theories fail, but when they succeed—*and still feel insufficient*. This is where discomfort enters. Not as confusion, but as a subtle misfit between explanation and experience.

In practice, reality often demands something that force-centric language struggles to express.

A bubble forms only when pressure is gentle. A fluid behaves differently depending on how quickly it is disturbed. Friction stabilizes motion instead of merely opposing it. Chaos appears when systems are rushed, not merely when they are strong.

None of these observations contradict theory. And yet, they do not feel fully explained by it either. The equations work, but the *story* feels incomplete.

This discomfort arises because lived phenomena are conditional, while our explanations are often absolute. Force-based descriptions emphasize magnitude—how much, how strong, how large—while experience repeatedly reveals that *rate*, *timing*, and *capacity* matter just as much, if not more. When these factors dominate behavior, intuition senses that something essential is missing from the causal picture.

The problem is not that theory ignores these factors. It often includes them implicitly, buried inside coefficients, boundary conditions, or correction terms. But when such elements are treated as secondary adjustments rather than primary causes, explanation begins downstream of reality's true ignition point.

As a result, many systems feel alive in ways theory cannot comfortably articulate. They appear responsive rather than reactive, sensitive rather than rigid, adaptive rather than mechanical. Stability feels negotiated, not enforced. Motion feels released, not imposed. Breakdown feels like overload, not opposition.

When explanation begins with force, these qualities appear accidental. They are treated as exceptions, complications, or nonlinearities—features to be managed rather than signals to be understood. The discomfort grows because experience insists that these features are central, not peripheral.

This is why skill matters more than strength in many real-world interactions. Why slow often succeeds where fast fails. Why coherence feels fragile yet robust. Why intuition resists explanations that technically work but conceptually lag behind lived reality.

Importantly, this discomfort is not a failure of education or intelligence. It is not the reader "not getting it." It is the mind recognizing a causal inversion—a sense that explanation has started after the crucial event has already occurred.

Reality does not begin with force. It begins with a condition that *demands* force-like behavior later.

When theory meets reality and hesitates, it is because explanation has arrived late. The discomfort is the mind's way of signaling that the true beginning lies upstream, where imbalance first appears and resolution becomes inevitable.

This discomfort is not something to suppress. It is the doorway.

Once acknowledged, it prepares the reader to ask a different question—not *how does motion behave*, but *what makes motion necessary at all.*

1.5 — The Cost of Explaining from the Wrong End

Explaining the world from the wrong starting point does not stop understanding—it quietly distorts it. The cost is not immediately visible because predictions still succeed and technologies still function. But over time, the explanatory framework accumulates strain. Concepts multiply. Exceptions grow. Intuition fragments. What should feel unified begins to feel patched together.

When force is treated as the origin of change, several consequences follow naturally—and almost unnoticed.

First, coherence appears accidental. Stability is explained as balance, but balance itself is left unexplained. Why systems prefer certain configurations, why they resist some disturbances but yield to others, why they stabilize dynamically instead of collapsing—all of this is pushed into secondary reasoning. Order becomes something that must be justified after the fact, rather than something that emerges naturally from how imbalance resolves.

Second, chaos becomes mysterious. If force is the cause, then chaos looks like a failure of control or predictability. Turbulence, breakdown, noise, and loss are treated as complications layered on top of otherwise clean laws. Entire mathematical frameworks are built to manage chaos, yet its origin remains framed as an exception rather than a consequence of overwhelming misalignment. Chaos feels external to order, instead of its natural limit.

Third, friction and dissipation feel artificial. Losses are introduced as opposing terms, correction factors, or empirical constants. Energy conservation is preserved formally, but the *experience* of loss—why motion settles, why systems calm down, why excess disappears into heat—feels bolted on rather than intrinsic. Dissipation becomes something to account for, not something to understand.

Fourth, the observer becomes inflated. When causality is unclear at the foundation, observation begins to carry explanatory weight it does not deserve. Meaning, collapse, choice, and agency drift toward the observer simply because the framework lacks a clear account of how resolution occurs independently. Mystery migrates from nature into perception, and the observer is quietly elevated from participant to arbiter.

Finally, domains fracture. Physics, chemistry, biology, psychology, and social dynamics start requiring different explanatory languages, even though they deal with the same reality. Each domain invents its own causes—forces here, drives there, incentives elsewhere—because the underlying condition driving change was never unified. What should scale smoothly instead splinters conceptually.

None of these outcomes imply that force-based physics is wrong. They reveal that it is partial. It explains regulation without ignition, response without invitation, motion without necessity. It tells us how systems behave once they are already dynamic, but not why dynamism must arise in the first place.

The deeper cost is subtle but profound: the loss of causal simplicity. When the true beginning is missed, explanation must work harder downstream. Concepts multiply where one would suffice. Mystery grows where clarity is possible. Intuition struggles to stay aligned across scales.

Reordering causality does not discard what we know. It relieves it.

When imbalance is placed first, force regains its rightful role as consequence rather than cause. Chaos becomes understandable.

Dissipation becomes natural. The observer returns to scale. Domains reconnect.

The price of explaining from the wrong end is not failure. It is unnecessary complexity.

And complexity, once normalized, is remarkably difficult to question—until a simpler beginning finally comes into view.

Turning the Map Without Burning It

This chapter has not argued against physics, nor has it challenged the power of its laws. Nothing discussed here requires abandoning equations, predictions, or the immense body of knowledge that has been built with care and precision. What has been questioned is far quieter—and far more consequential.

It is the starting point.

For a long time, we have explained the world from the middle. We have begun where calculation becomes possible, where quantities can be related, balanced, and conserved. From there, we have worked outward with remarkable success. But beginnings matter. When causality is placed downstream, explanation inevitably feels strained when it encounters living, flowing, rate-sensitive reality.

The discomfort explored in this chapter is not an attack on science. It is a diagnostic. It tells us that explanation is arriving late—that something essential has already happened by the time force, motion, or equilibrium enter the story. That something is not mysterious, hidden, or metaphysical. It is simply unnamed.

This chapter has prepared the ground by loosening one assumption: that force must be the first word in the language of nature. It has suggested that imbalance may precede force, that equations may describe outcomes rather than causes, and that much of our conceptual difficulty arises from this quiet inversion.

Nothing needs to be torn down. The map only needs to be rotated.

In the next chapter, we will strip the problem even further. We will look at energy itself—undeniably present, rigorously conserved, and yet often inert. We will ask a deceptively simple question: if energy exists everywhere, why does nothing happen everywhere?

That question leads naturally to what force cannot explain alone.

Before motion can be calculated, something must already be unsettled.

With that unsettled condition brought into focus, the real beginning of dynamism can finally be approached.

Chapter 2 — Energy Exists, Nothing Happens

When Presence Does Not Become Action

Energy is one of the most confidently spoken words in science. It is everywhere, conserved across all processes, measurable, transferable, and foundational. We are taught early that energy is what makes things happen. And yet, when we look carefully at the world around us, a quiet contradiction appears.

Energy exists everywhere—and most things do nothing.

Mountains sit unmoved for centuries. A book rests on a table without complaint. A charged system remains silent. A compressed spring holds perfectly still. Vast energy is present, accounted for, conserved—and yet, nothing unfolds. No motion, no change, no manifestation. The world does not seem eager to act simply because it can.

This observation is so ordinary that it is rarely questioned. We accept stillness as normal, even though it appears to contradict the idea that energy is the engine of activity. If energy were sufficient to produce motion, the universe would never pause. Everything would constantly transform. But it does not.

Static systems feel strangely "dead," not because they lack energy, but because nothing within them demands resolution. They feel complete. Finished. Closed. This feeling is not poetic; it is physical. When nothing is misaligned, there is no reason for anything to unfold.

This creates an important distinction that is often blurred: the difference between capacity and necessity. Energy represents capacity—the ability for change to occur. But capacity alone does not compel action. Having the ability to move is not the same as needing to move.

We sense this intuitively in everyday life. Possessing strength does not require exertion. Having resources does not force expenditure. Readiness does not imply urgency. In the same way, energy does not demand expression simply because it exists.

The confusion arises because many of our explanations quietly assume that energy wants to act, that motion is its natural state, and that stillness is something that must be justified. But reality suggests the opposite. Stillness is natural when nothing is unsettled. Motion appears only when something *must* change.

This chapter begins with that silence.

Before introducing delta, before naming imbalance, before reframing motion, it is essential to recognize this simple fact: energy can exist indefinitely without becoming dynamic. Something else must invite change. Something must disturb completion, not violently, but decisively enough to make resolution unavoidable.

Understanding this difference is the first step in turning explanation right-side up.

Energy is present everywhere. But presence alone does not make the world move.

2.1 — Energy as Capacity, Not Compulsion

At its most fundamental level, energy represents capacity. It is the ability to do work, to support change, to enable transformation if conditions demand it. This definition is precise, widely accepted, and quietly revealing. Capacity does not imply action. It implies readiness—nothing more.

A system may possess enormous energy and remain completely inactive. This is not a contradiction; it is the natural state of things. Energy does not carry intention. It does not seek expression. It does not urge motion simply because it exists. It waits.

This is why conservation laws are so stable and so silent. Energy persists through every process unchanged in total amount, indifferent to whether anything happens or not. The universe is full of energy that is doing nothing at all. That inactivity is not a failure of energy—it is its default condition when no resolution is required.

The habit of associating energy directly with action comes from conflating *ability* with *necessity*. In daily language, we speak of "having energy" as being active, motivated, or restless. But physically, energy has no such urgency. A fully charged battery on a shelf does not act. A massive object at rest does not move. A pressurized system in perfect balance remains silent.

Energy enables change, but it does not initiate it.

This distinction matters because many explanations quietly assume that energy must express itself. Motion is then treated as inevitable, and stillness becomes something that needs explanation. In reality, it is the other way around. Stillness is what we should expect when nothing is misaligned. Motion requires justification.

Once energy is understood as capacity rather than compulsion, a subtle shift occurs. The question is no longer "why is this system still?" but "what would make it unable to remain still?" That question points away from energy itself and toward conditions that disrupt completeness.

Energy does not push reality forward. It merely allows reality to move *if something else demands it*.

Recognizing this is essential, because it clears the conceptual space needed to identify what actually ignites change. Until energy is freed from the burden of causation, the true driver of dynamism remains hidden.

Energy provides the possibility. Something else provides the invitation.

2.2 — Dormant Energy and the Silence of Static Systems

When a system is static, it often feels uneventful, even lifeless. Nothing moves, nothing changes, nothing draws attention. Yet this apparent emptiness is misleading. Static systems are rarely empty of energy. They are simply complete.

A book resting on a table is not devoid of energy. Gravitational energy is present, internal molecular energy persists, thermal motion continues at microscopic scales. And still, the system remains unchanged. Nothing unfolds because nothing needs to. Every tendency toward change is already satisfied by the system's configuration.

This is what dormancy looks like in physical terms: energy without demand.

Static equilibrium is often misunderstood as a delicate balance of opposing forces, as if the system were constantly struggling to remain still. In reality, well-settled systems do not struggle at all. They are not resisting motion; they are free from the necessity of motion. All potential pathways of change are either closed or already resolved.

This is why static systems feel "dead." Not because energy is absent, but because resolution has finished its work. There is no unfinished business. No gradient calling for rearrangement. No misalignment insisting on release.

Importantly, dormancy is not fragility. A static system can be extraordinarily robust. Mountains remain still not because they are weak, but because the conditions required to disturb them are extreme. Their silence is not emptiness; it is stability achieved through completion.

This perspective reframes stillness entirely. Stillness is not the absence of physics. It is physics that has nothing left to do.

When motion does occur, it is tempting to say that energy has been "activated." But energy was always there. What changed was not the energy, but the condition of the system. Something that had been settled became unsettled. A gradient appeared where none existed before. Dormancy gave way to demand.

Understanding this resolves a long-standing confusion. We often search for the cause of motion within energy itself, assuming that something must have been added or released. But motion does not require additional energy; it requires a reason for energy to reorganize.

Static systems teach us a crucial lesson: energy alone does not animate the world.

They show us that silence is not mysterious. It is the natural outcome of completion. Only when that completion is disturbed—when something becomes misaligned—does the quiet break, and dynamics begin.

Dormant energy is not waiting impatiently. It is simply at rest, because nothing is asking it to move.

2.3 — Why Potential Alone Does Not Create Motion

Potential energy is often spoken of as if it were motion waiting to happen. A raised object "wants" to fall. A stretched spring "wants" to contract. A charged system "wants" to discharge. This language is convenient, but it quietly misleads. Potential energy does not want anything. It merely permits something to happen *if* conditions require it.

A system can hold enormous potential energy and remain perfectly still indefinitely. A boulder perched on solid ground at height contains gravitational potential, yet it does not move. A compressed spring locked in place remains silent. A chemical system can store vast energetic capacity without reacting. In each case, energy is present, quantified, and conserved—yet motion does not follow.

This reveals a crucial truth: potential is conditional, not directive.

Potential energy describes what *could* happen under a change in conditions. It does not describe what *must* happen. Motion only appears when the system is placed in a situation where remaining unchanged is no longer possible. Until then, potential energy is simply stored permission.

The confusion arises because many textbook examples introduce potential energy and release simultaneously. A block is raised and then immediately allowed to fall. A spring is compressed and then released. A capacitor is charged and then discharged. The moment of misalignment is hidden inside the setup, so potential energy appears to be the cause of motion, when in fact it is only the context that makes motion possible.

What actually initiates motion in these cases is not the existence of potential, but the creation of a gradient—a situation in which the system cannot remain complete without reconfiguration. The moment a constraint is removed, a balance broken, or a pathway opened, potential energy gains relevance. Without that change, it remains inert.

This distinction becomes clearer in systems where potential energy is abundant but motion is rare. Large-scale geological formations, stable atomic structures, and long-lived astrophysical systems all store energy across immense timescales without dynamic expression. Their longevity is not evidence of low energy, but of low necessity.

Potential energy answers the question, *What could happen if conditions change?* It does not answer the question, *Why does anything happen now?*

By treating potential as an automatic precursor to motion, explanation again begins too late. The true trigger lies upstream, in the moment when completeness is broken and resolution becomes unavoidable. Potential energy then participates in the process—but it does not initiate it.

This is why motion cannot be explained by energy magnitude alone. Systems with less energy can be highly dynamic, while systems with vast energy can remain silent. What matters is not how much energy exists, but whether the system is allowed—or forced—to reorganize.

Potential energy is readiness, not impulse. It waits for something else to make change necessary.

That "something else" is not force, and not energy itself. It is the condition that turns possibility into process—the condition this book calls delta.

2.4 — The Illusion of Force as the Ignition

Force is often introduced as the moment where motion begins. A push is applied, an object accelerates, and the story appears complete. This framing is intuitive, visually compelling, and deeply ingrained. It gives the impression that force *creates* motion. But this impression survives only because something crucial has already been assumed and quietly hidden.

Force never appears in isolation.

Every instance where force is invoked already contains a prior condition—an imbalance, a gradient, a misfit in configuration—that makes force relevant in the first place. What we call force is not the ignition of motion; it is the formal expression of a system that can no longer remain as it is.

Consider what "net force" actually means. It is not an entity acting independently; it is a statement that something is unequal. Unequal pressures, unequal potentials, unequal constraints. Net force is simply imbalance written in mathematical language. By starting explanations with force, we begin after the real event has already occurred.

This is why force feels so effective and yet so unsatisfying as a first cause. It works because it captures how systems respond once imbalance exists. It fails as an origin because it cannot explain why imbalance arose, or why resolution had to begin at that moment.

Force-based explanations also obscure an important experiential fact: motion does not feel imposed; it feels released. When a constraint is removed, motion unfolds naturally. When a system is gently allowed to reorganize, it does not struggle—it flows. This sense of release is not poetic intuition; it reflects the underlying reality that motion follows from necessity, not from coercion.

The illusion of force as ignition persists because force is where calculation becomes easy. It is the point at which variables can be related cleanly. But ease of calculation is not evidence of causal primacy. It merely marks the point where mathematics becomes comfortable.

Once force is mistaken for ignition, several confusions follow. Resistance appears as opposition rather than regulation. Friction appears as an enemy of motion rather than a resolver of excess. Stability appears fragile, as if it must constantly defend itself against forces, rather than robust, achieved through completion.

Most importantly, force-centric thinking disguises the role of conditions. Rate, timing, boundary readiness, and internal capacity all become secondary modifiers instead of central determinants. Yet in real systems, these factors decide everything.

Force does not start motion. It formalizes motion once the system has already been unsettled.

By mistaking expression for origin, explanation again begins too late. The true ignition lies upstream, in the moment when a system loses the ability to remain complete without change. Force then appears—not as cause, but as consequence.

Recognizing this does not weaken force-based laws. It relocates them.

Force belongs to the language of response, not the language of initiation. And once this is seen, the search for the true ignition becomes unavoidable.

2.5 — Delta as the Missing Ignition Key

If energy is capacity and force is response, then something else must answer the simplest unanswered question: what makes capacity become active at a particular moment? That missing element is delta.

Delta is not an extra ingredient added to a system. It is a condition—the presence of misalignment that makes remaining unchanged impossible. Until delta appears, energy remains dormant and force remains undefined. Once delta exists, resolution becomes unavoidable.

This is why motion does not begin everywhere energy exists, but only where something is unsettled.

Delta is often confused with difference, disturbance, or noise. It is none of these by necessity. A difference can exist without consequence. A disturbance can be absorbed without change. Noise can fluctuate without direction. Delta, by contrast, is a misalignment that cannot be ignored by the system's internal organization. It is the point where completeness breaks.

When delta appears, energy does not suddenly awaken; it is recruited. Stored capacity becomes resolving energy. The system begins to reorganize, not because it wants to, but because it no longer has the option to remain as it was. Motion, flow, deformation, reconfiguration—all follow from this single fact.

This reframes ignition entirely. Motion does not start because energy is released. Energy is released because delta demands resolution. Force does not initiate change. Force appears as the measurable expression of how that resolution proceeds.

Seen this way, delta explains several persistent puzzles at once.

It explains why rate matters more than magnitude. A large delta introduced slowly can be resolved coherently, while a smaller delta introduced too quickly can cause breakdown. It explains why gentleness often succeeds where strength fails. It explains why systems have thresholds, beyond which behavior changes qualitatively rather than quantitatively.

Most importantly, delta explains why nothing happens until it must.

A static system is not waiting for energy; it is waiting for delta. A dynamic system is not driven by force; it is resolving misalignment. Once delta is fully resolved, motion ceases naturally, without opposition or exhaustion. Silence returns—not because energy is gone, but because resolution is complete.

This is the ignition key that force-centric explanations quietly skip. By beginning with force, they begin after delta has already done its work. By restoring delta to its rightful place, causality becomes continuous rather than fragmented.

Energy provides the capacity. Force describes the response. Delta makes change unavoidable.

With this ignition identified, the next step becomes clear: delta must be understood not as a quantity to be inserted into equations, but as a condition that precedes them. That task belongs to the next chapter, where delta is examined in its own right—not as a variable, but as the true beginning of dynamism.

2.6 — Resolving Energy: When Energy Becomes Time

When delta appears and resolution begins, energy undergoes a subtle but decisive transformation. It does not change in quantity, nor does it acquire new properties. What changes is its mode of existence. Energy stops being merely stored capacity and becomes process. In that transition, time emerges.

Time is not something added to energy. Time is how resolving energy is experienced.

As long as energy remains dormant—perfectly balanced, fully settled—there is no internal reference for before and after. Nothing unfolds. No sequence asserts itself. The system is complete, and completion has no duration. The moment delta disrupts that completeness, resolution introduces direction. That direction is what we experience as time.

This is why time feels inseparable from change. It does not flow independently; it appears wherever resolution is underway. When resolution accelerates, time feels dense and urgent. When resolution slows, time stretches and thins. When resolution completes, time effectively disappears from experience.

In this sense, energy becomes time only when it is resolving.

This reframing clarifies why static systems feel timeless. They are not frozen in time; they are finished with it. No internal reorganization is occurring that could mark progression. By contrast, dynamic systems generate time continuously because resolution is ongoing. Every step of reconfiguration establishes a "next," and the sequence of these nexts is what we call temporal flow.

Motion, flow, deformation, learning, growth—these are not separate phenomena layered on top of time. They *are* time made visible. Each is a particular way in which energy resolves delta across a system's structure. Time does not govern these processes; it is their shadow.

This perspective also explains why time is relative without invoking abstraction. Different systems resolve delta at different rates and through different pathways. A fast-resolving system experiences a compressed sense of progression. A slow-resolving system experiences extended continuity. There is no universal clock ticking behind reality—only resolution unfolding where it must.

Understanding time this way removes its mystery without diminishing its importance. Time is not an external container. It is not a fundamental dimension in which energy moves. It is the measure of how misalignment is being resolved.

Energy exists whether time is felt or not. Delta introduces necessity. Resolution creates sequence.

And sequence, once experienced, is time.

This completes the shift begun in this chapter. Energy is no longer mistaken for motion. Force is no longer mistaken for ignition. Change is no longer mysterious. With delta identified as the condition that makes resolution unavoidable, and with time recognized as the experience of that resolution, the groundwork is laid for a precise definition of delta itself.

The next chapter turns fully toward that task—examining delta not as a number to be calculated, but as the condition that brings reality into motion at every scale.

From Capacity to Invitation

This chapter has stripped motion of its false urgency. It has shown that energy, no matter how abundant, does not rush to express itself. It waits. It remains silent until something makes silence untenable. Stillness, far from being an absence, is revealed as completion—energy with nothing left to resolve.

By separating capacity from necessity, we have quietly removed one of the most persistent confusions in how change is explained. Energy does not initiate dynamics. Potential does not compel motion. Force does not ignite activity. All of these enter the picture only after a more basic condition has already come into existence.

That condition is misalignment.

When something becomes unsettled—when a system can no longer remain complete without reconfiguration—energy is no longer dormant. It becomes resolving energy. In that act of resolution, motion appears, time becomes perceptible, and structure begins to unfold. Change is not imposed; it becomes unavoidable.

This reframing does not diminish energy. On the contrary, it restores its dignity by freeing it from a role it was never meant to play. Energy is not restless. It is capable. It does not demand expression. It allows expression when something else calls for it.

That call is delta.

Up to this point, delta has appeared only as an implication—an unnamed necessity that precedes force, motion, and time. In the next chapter, delta steps into the foreground. Not as a variable to be inserted into equations, but as a condition that must be recognized. A condition without which no equation would ever be needed, and with which every dynamic process becomes intelligible.

If previous chapter turned the explanatory map around, and this chapter identified what does *not* ignite reality, then next chapter takes the next unavoidable step: it defines what does.

Energy is capacity. Delta is invitation.

With that invitation clearly in view, the true beginning of dynamism can finally be named.

Chapter 3 — Delta: Not a Quantity, but a Condition

The Mistake of Measuring Too Early

The moment a new idea appears, our instinct is to measure it. We ask how much, how fast, how strong, how large. Numbers feel safe. They give shape to uncertainty and promise control. In science especially, measurement has become synonymous with understanding. What can be quantified is considered real; what cannot is treated with suspicion.

But measurement always arrives after something has already happened.

Before a quantity can change, a condition must exist that allows change to begin. Before a variable can vary, something must make variation necessary. When we rush to measure too early, we risk mistaking the *expression* of a process for its *origin*.

Delta has suffered precisely this fate.

In most contexts, delta is introduced as a difference—between two values, two states, two moments. It is written as a symbol, inserted into equations, and treated as a calculable quantity. This makes delta appear secondary, derived, and dependent on prior definitions. But in doing so, something essential is lost. The deeper role of delta as a condition disappears behind arithmetic.

A difference alone does not guarantee change. Two states may differ and yet coexist without consequence. Two values may be unequal and still remain stable. Measurement can capture disparity, but it cannot tell us whether that disparity matters. Change begins not when numbers differ, but when a system can no longer remain aligned in the presence of that difference.

This is where premature measurement misleads.

By focusing on how much changes, we skip over the more fundamental question: *why does change have to occur at all?* That "why" cannot be answered by numbers, because it precedes them. It belongs to the realm of conditions, not quantities.

Delta, in its deepest sense, is not something to be calculated first. It is something to be recognized. It is the moment when compatibility breaks, when completion becomes impossible, when remaining as one is becomes more costly than becoming something else. Only after this condition exists does calculation become meaningful.

This chapter asks the reader to pause before measuring. To resist the urge to quantify immediately. To look instead for what must already be true before any equation can speak.

Delta is not introduced here as a variable to be solved. It is introduced as the reason solving ever becomes necessary.

Until that distinction is clear, every explanation begins too late.

3.1 — Delta Is Misalignment, Not Difference

Difference is easy to notice. Two values are unequal. Two states are not the same. One position is higher, another lower. Difference is descriptive, passive, and often harmless. It can exist indefinitely without demanding anything from the system in which it appears.

Delta is something else entirely.

Delta is misalignment—a condition in which parts of a system can no longer coexist without adjustment. It is not simply that things are different, but that their difference matters. It matters because the system cannot remain complete while accommodating it.

Two objects at different temperatures represent a difference. But until they are allowed to exchange energy, nothing happens.

Two regions at different heights represent a difference. But until a pathway exists, no motion follows.

Two ideas in the mind may contradict each other. But until action or choice is required, tension does not arise.

Difference describes contrast. Delta describes incompatibility.

This distinction is subtle but decisive. A system can tolerate many differences without consequence. Nature is full of gradients that remain stable because they are supported by structure, constraints, or alignment. Delta appears only when that support fails—when the system can no longer absorb difference without reorganizing.

This is why delta cannot be reduced to a numerical gap. A large numerical difference may produce no delta at all, while a small one can trigger dramatic change. What matters is not magnitude, but context—the system's capacity to remain aligned in the presence of that difference.

Delta is therefore relational, not absolute. It depends on structure, boundaries, timing, and internal coherence. The same difference can be inert in one situation and explosive in another. The difference did not change; the alignment did.

Seen this way, delta is not an object or a property. It is a state of tension between what is and what can be sustained. It marks the point where compatibility ends and resolution must begin.

This is why delta always precedes action. Until misalignment exists, there is no reason for anything to happen. Once misalignment appears, something must give—not because of force, not because of energy, but because the system cannot remain whole otherwise.

Difference can be observed. Delta must be resolved.

By recognizing delta as misalignment rather than difference, we shift explanation to where it truly begins. Not at motion, not at force, not at numbers—but at the moment when remaining unchanged becomes impossible.

Only then does change become inevitable.

3.2 — Why Delta Precedes Force, Motion, and Time

If delta is misalignment, then it must logically exist before anything responds to it. Force, motion, and time are not initiators; they are expressions. They appear only after a system has crossed the threshold where remaining unchanged is no longer possible.

This ordering is crucial.

Force does not announce itself in a settled system. Motion does not arise in completion. Time does not register where nothing unfolds. All three require a prior condition—a state that demands resolution. That state is delta.

Consider force first. What we call force is always a response to incompatibility: unequal pressures, uneven constraints, unbalanced configurations. "Net force" is not a thing acting on a system; it is a declaration that alignment has failed. By the time force can be defined or measured, delta has already done its work. The system is no longer whole, and response is unavoidable.

Motion follows the same logic. Motion is not an act imposed on matter; it is misalignment resolving through available pathways. A system does not move because it is instructed to do so. It moves because it cannot remain as it is. Motion is delta unfolding, not delta appearing.

Time, too, arrives late. Time is not a background in which events occur; it is the experience of sequence generated by resolution. Where nothing resolves, nothing progresses. Static systems do not feel slow; they feel timeless. Only when delta introduces necessity does resolution create a before and after. Time is not what allows change—it is what change produces.

This reverses a deeply ingrained intuition. We often imagine time flowing first, enabling motion, which then requires force. In reality, the chain runs the other way. Delta appears. Resolution begins. Motion unfolds. Sequence accumulates. Time is experienced.

This ordering explains why force-centric explanations feel complete yet unsatisfying. They begin at response, not at necessity. They tell us how systems behave once delta exists, but remain silent on why delta must exist at all. By placing delta first, causality becomes continuous rather than fragmented.

Nothing dynamic can exist without delta. Nothing responsive can act without misalignment. Nothing temporal can be felt without resolution.

Delta is therefore not an optional abstraction layered onto physics. It is the logical prerequisite for every phenomenon that follows. Before force can be measured, before motion can be observed, before time can be counted, something must already be wrong enough to demand change.

That "wrongness" is not error or failure. It is the precise condition that makes reality dynamic.

Delta does not compete with force, motion, or time. It precedes them—quietly, necessarily, and always.

3.3 — Delta Without Resolution: Tension

When delta appears but cannot resolve, the system does not become dynamic—it becomes strained. This state is commonly misunderstood as action or instability, but it is neither. It is postponed change. Delta exists, necessity is present, yet pathways for resolution are blocked, delayed, or insufficient. What remains is tension.

Tension is not motion. Tension is unresolved demand.

Physically, this shows up as stress in materials, pressure in confined fluids, charge separation without discharge, thermal gradients without flow. Psychologically, it appears as anxiety, anticipation, cognitive dissonance. Socially, it manifests as friction without reform, conflict without movement. In every domain, the pattern is the same: misalignment exists, but resolution is constrained.

This is why tension feels heavy and urgent. It carries direction without release. Energy is available, delta is present, but alignment cannot proceed. The system is forced to *hold* incompatibility. That holding has limits.

Unresolved delta accumulates. It does not disappear. It stores itself as strain, compression, pressure, or stress—waiting for conditions to change. This waiting is not passive; it is precarious. The longer resolution is postponed, the more sensitive the system becomes to small perturbations. A minor change in boundary conditions can trigger sudden release, often appearing dramatic or violent precisely because resolution has been delayed.

This is the origin of thresholds.

A bridge under load does not fail because force suddenly appeared; it fails because delta exceeded the structure's capacity to keep holding without resolving. A fluid confined under pressure does not erupt randomly; it erupts when containment can no longer postpone reconfiguration. A mind under prolonged contradiction does not break without cause; it breaks when internal alignment cannot be maintained any longer.

Tension, then, is not instability—it is stability under duress. It is the system doing everything it can to remain unchanged in the presence of misalignment. This effort is finite. Every system has an alignment capacity beyond which postponement becomes impossible.

Understanding tension this way removes its mystery. It is not an opposing force, not a buildup of aggression, not an abstract discomfort. It is delta without a viable path forward. It is necessity waiting for permission.

Once permission arrives—through a pathway opening, a boundary weakening, or a rate adjusting—resolution begins, and tension vanishes. Not because energy was spent, but because incompatibility was finally allowed to reorganize.

Delta without resolution does not produce motion. It produces strain.

And strain is not failure. It is the system signaling that change is required, but not yet allowed.

Recognizing this is crucial, because it explains why systems can remain apparently stable for long periods and then change abruptly. The abruptness is not sudden delta; it is sudden resolution of delta that had been present all along.

Tension is the quiet proof that delta exists.

3.4 — Resolution Without Delta: Silence

If delta without resolution creates tension, then the opposite condition—resolution without delta—creates silence. This silence is often mistaken for emptiness, inactivity, or absence. In reality, it is the most complete state a system can occupy.

Silence is not nothingness. Silence is nothing left to resolve.

When a system is fully aligned, there is no incompatibility within it that demands change. Energy is present, structure exists, capacity remains—but there is no necessity. Nothing pulls, nothing pushes, nothing unfolds. The system does not resist motion; it simply has no reason to move.

This is why equilibrium feels uneventful.

In a perfectly aligned system, forces do not cancel in a dramatic standoff. They are irrelevant. Motion does not slow down gradually; it never begins. Time does not drag; it loses meaning. The system is not frozen—it is finished, at least for the moment.

We often misinterpret silence as passivity because our intuition associates activity with existence. But nature is under no obligation to remain expressive. A stable atom, a resting object, a calm fluid, a settled mind—all are examples of systems where resolution has already occurred. Their stillness is not fragile; it is earned.

This explains why silence is robust.

A system without delta does not need protection from disturbance; it simply absorbs it until misalignment truly appears. Small perturbations vanish because there is nothing for them to resolve into. Without delta, fluctuations have no direction, no consequence, no memory.

This is also why silence feels timeless. Where nothing resolves, sequence does not accumulate. There is no internal marker for before and after. From the system's own perspective, nothing progresses because nothing needs to.

Importantly, silence does not oppose change. It merely withholds it until conditions require otherwise. The moment misalignment appears—through boundary shifts, external coupling, or internal rearrangement—silence gives way effortlessly. No struggle, no delay. Resolution begins naturally because it must.

Understanding silence this way corrects a deep misunderstanding. Stillness is not the absence of physics. It is physics that has completed its task. The laws have nothing left to do.

Delta is the only thing that can break silence. Resolution is the only thing that can end tension.

Without delta, there is no story to tell. And without story, there is silence.

This silence is not mysterious. It is the most honest expression of alignment.

3.5 — Delta as a Condition, Not a Variable

At this point, it becomes necessary to draw a firm boundary between two very different ways of thinking. One treats delta as a variable—a number to be calculated, adjusted, optimized. The other recognizes delta as a condition—a state of compatibility or incompatibility that exists before any calculation becomes meaningful.

Most confusion about delta arises from collapsing these two.

When delta is treated as a variable, it is assumed to behave like other quantities: something that can be increased, decreased, substituted, or solved for. This makes delta appear dependent on equations, units, and measurement frameworks. In this view, delta becomes secondary—an output of formalism rather than the reason formalism exists at all.

But conditions are not variables.

A condition does not need to be measured to be real. It needs to be present. Dryness, tension, readiness, instability, saturation—these are conditions. They determine what is possible long before any numerical description is attempted. Delta belongs to this category. It is not "how much difference exists," but whether the system can remain whole in the presence of what exists.

This is why delta must be recognized, not computed.

Equations implicitly assume delta everywhere. Pressure gradients, potential differences, concentration gradients, stress fields—these are all mathematical expressions of systems that are already misaligned. The equations do not create delta; they only describe how resolution proceeds once delta is given. By the time symbols appear on paper, the condition has already been met.

Treating delta as a variable reverses causality again. It suggests that delta emerges from equations, when in fact equations emerge because delta has already made change unavoidable. The system demands description because it is no longer complete.

This distinction also explains why delta resists universal quantification. The same numerical difference can be irrelevant in one system and decisive in another. A small delta can collapse one structure and leave another untouched. No single scale captures delta's importance, because importance is not numerical—it is contextual.

Conditions depend on structure, boundaries, timing, and internal organization. Delta inherits all of these dependencies. It cannot be reduced to a scalar without losing its essence. Once reduced, it becomes just another difference—stripped of necessity, stripped of urgency, stripped of meaning.

Recognizing delta as a condition simplifies explanation instead of complicating it. It removes the need to invent special forces, hidden variables, or observer-dependent triggers. Change happens because remaining unchanged is no longer an option. That fact requires no equation to be true.

Only after delta is recognized as a condition does it make sense to ask how resolution proceeds, at what rate, through which pathways, and with what consequences. Variables belong to that later stage. Delta belongs to the beginning.

Delta is not something you plug into reality. It is something reality responds to.

By restoring delta to its rightful place—as condition rather than quantity—we finally place explanation where it belongs: before motion, before force, before time, and before calculation.

What follows in the next section is the unavoidable conclusion of this recognition: delta is not confined to any domain or scale. If it is truly a condition of dynamism, then it must appear everywhere change appears.

3.6 — The Universality of Delta

If delta is truly a condition rather than a quantity, then it cannot belong to any single domain of physics, scale of reality, or category of explanation. Conditions do not specialize. They govern. Delta must therefore appear everywhere that change, motion, organization, or collapse appears—otherwise it would be a local artifact, not a foundational principle.

This is precisely what observation reveals.

In physical systems, delta appears as misalignment in pressure, temperature, concentration, configuration, or constraint. Fluids flow not because they are pushed, but because gradients make stillness unsustainable. Friction dissipates not to oppose motion, but to resolve excess misalignment at interfaces. Structures fail not when forces appear, but when alignment capacity is exceeded.

In biological systems, delta governs life itself. Metabolism exists because internal chemical states cannot remain aligned without continuous resolution. Growth occurs where internal and external conditions mismatch in a way that demands adaptation. Death is not the loss of energy, but the loss of the system's ability to resolve internal deltas coherently.

In cognitive systems, delta appears as contradiction, uncertainty, unmet intention, or imbalance between perception and expectation. Thought does not arise in silence; it arises where mental alignment fails. Learning occurs when delta is resolved. Anxiety persists where delta cannot resolve. Insight appears when resolution reorganizes the internal landscape.

In social systems, delta manifests as inequality, tension, unmet need, or incompatible structures. Revolutions do not occur because force suddenly increases, but because misalignment exceeds the system's capacity to contain it. Reform succeeds when delta is resolved gradually; collapse follows when it is postponed too long.

Across all these domains, the pattern is invariant:

- No delta → no dynamism
- Delta unresolved → tension
- Delta resolving coherently → flow
- Delta overwhelming resolution → chaos

Scale changes nothing. Context changes expression, not principle.

This universality is not metaphorical. It is structural. Delta does not migrate from physics to psychology as an analogy; it remains the same condition appearing through different organizations of matter, energy, and information. The language changes, the phenomena change, but the causal role does not.

This is why delta can unify explanations without flattening them. It does not reduce complexity; it organizes it. Each domain retains its specificity, but all share the same ignition condition. The world does not need separate causes for motion, life, mind, and society. It needs one condition that explains why any of them must change at all.

Recognizing delta's universality removes the need for privileged explanations. There is no special realm where different rules apply. There is no domain where change begins for a different reason. Everywhere something happens, delta has already appeared.

This realization carries an important implication: delta is not an abstraction invented to connect disparate fields. It is the common reality those fields have been describing separately, without naming it.

Once delta is seen as universal, the remaining task becomes unavoidable. If delta governs all dynamism, then its behavior must have regimes, limits, and thresholds. It must explain why coherence appears sometimes and chaos at others. It must account for timing, scale, and stability.

That task begins in the next part of the book, where delta is no longer simply identified, but examined in motion—through its regimes, its thresholds, and its capacity to create order or destroy it.

Delta has now been named. What remains is to understand how it behaves.

Naming the True Beginning

This chapter completes a reversal that has been quietly underway since the opening pages of the book. Delta has been stripped of its usual disguises—difference, disturbance, fluctuation, noise—and restored to what it actually is: a condition of misalignment that makes change unavoidable.

By now, several distinctions should feel settled rather than argued.

Delta is not a number.

Delta is not a force.

Delta is not motion, time, or energy in disguise.

Delta is the moment a system can no longer remain whole without reorganization.

Seen this way, force regains clarity as response, motion as unfolding, and time as experienced sequence. Tension is revealed as postponed resolution. Silence is revealed as completion. Nothing mystical is required, and nothing familiar is rejected—only reordered.

Most importantly, explanation has finally reached its true starting point.

Before this chapter, delta appeared as an implication—something sensed but unnamed. Here it has been defined carefully enough to stand on its own, without borrowing authority from equations or metaphors. It does not compete with existing laws. It precedes them. It explains why laws are needed at all.

This matters because once the beginning is correctly identified, everything downstream simplifies. The questions that follow are no longer vague or philosophical. They become precise and unavoidable:

- Why do some deltas resolve smoothly while others explode into chaos?

- Why does timing matter more than magnitude?
- Why do systems tolerate misalignment up to a point—and then fail abruptly?
- Why does coherence feel fragile yet persistent across nature and life?

These are not new questions. They have appeared repeatedly under different names: stability, turbulence, phase change, breakdown, learning, crisis. What has been missing is a unifying lens that explains them as variations of the same process.

Delta provides that lens.

With delta identified as the true beginning, the book now moves forward—not toward abstraction, but toward behavior. The next chapters will examine how delta behaves once it appears: how it resolves, how it accumulates, how it overwhelms, and how coherence emerges when resolution is aligned.

The foundation is now set. Nothing dynamic lies beneath delta.

Delta is not how much changes. Delta is why change becomes unavoidable.

From here on, the story is no longer about *what* exists, but about *how becoming unfolds.*

Interlude — When Explanation Finally Stops Chasing

Something subtle should have happened by now.

Not an insight that demands applause. Not a conclusion that feels triumphant. But a quiet relief—the sense that explanation has stopped running after reality and has finally stood where reality begins.

Until now, the book has not added complexity. It has removed urgency. Force no longer needs to explain everything. Energy no longer needs to act. Time no longer needs to flow on its own. The world has not become smaller—but calmer.

What has changed is not what you know, but where explanation starts.

When delta is recognized as the beginning, many long-standing puzzles lose their tension. Stillness no longer feels suspicious. Chaos no longer feels accidental. Friction no longer feels wasteful. Silence no longer needs justification. Nothing is "waiting to happen" unless something is misaligned.

This is not a philosophical move. It is a physical one.

Nature does not hesitate. Nature does not choose. Nature does not observe itself into existence.

Nature resolves.

Everything that appears dynamic—motion, flow, growth, collapse, learning, decay—is not an event layered onto reality. It is *resolution in progress*. The world does not move because it wants to, nor because it is instructed to. It moves because remaining unchanged is no longer possible.

At this point, you may notice something important: there has been no need to invoke intention, randomness, intelligence, or observers. And yet, nothing feels mechanical or dead. On the contrary, the world begins to feel *precise*, almost considerate—doing exactly what is required, no more, no less.

This is the moment where misunderstanding often creeps in.

Recognizing delta does not make reality passive. It makes it inevitable.

What follows now is not theory layered on top of theory. It is description catching up with experience. The next part of the book does not ask *why* change happens—that has already been answered. It asks *how* resolution proceeds once it begins.

From here on, the book stops correcting mistakes and starts describing motion as it actually unfolds.

This part established the true beginning of dynamism. Next part will trace its path.

Not through force. Not through intention. But through resolution—how delta unfolds, how coherence appears, how chaos erupts, and why timing matters more than magnitude.

The world does not act. It becomes.

And becoming has rules.

Part II — How Nature Actually Moves

With the origin of dynamism clarified, the task now shifts from *cause* to *process*. This part does not ask why change begins; it follows what happens once resolution is unavoidable. Here, nature is observed not as a collection of interacting objects, but as a continuous unfolding of alignment and misalignment across systems.

This part establishes a simple but powerful idea: *nature does not act, it resolves*. There are no decisions, no preferences, no intentions hidden in physical law. What appears as choice, direction, or behavior is the natural consequence of how delta is distributed and how resolution proceeds through available pathways.

Movement, flow, growth, decay, and transformation are shown to be expressions of the same underlying process. Resolution always moves in one direction—not toward a goal, but toward compatibility. Time emerges not as an external backdrop, but as the measure of this unfolding. Where resolution accelerates, time compresses. Where resolution slows, time stretches. Where resolution completes, time disappears from experience.

This part also introduces a crucial distinction between *regimes of delta*. Not all misalignment produces the same outcome. When delta overwhelms a system's capacity to resolve, coherence breaks and chaos appears. When delta is too small, change fragments into noise and discontinuous jumps. Only within a narrow, system-specific range does delta resolve smoothly, giving rise to stable structures, flow, and persistence. This regime is not accidental; it is where nature does its most precise work.

Finally, this part formalizes **Alignment Threshold Time (ATT)**—the idea that every system has a characteristic capacity to resolve delta over time. Instability arises not from delta alone, but from mismatch between delta and this capacity. What is often framed as impatience, inefficiency, or failure is revealed as a physical mismatch of rate and bandwidth. Patience, in this view, is not moral restraint; it is respect for resolution limits.

Together, these chapters shift the reader from static explanation to dynamic understanding. Nature is no longer something that *responds* to forces or *follows* laws. It is a system of continuous becoming, governed by the exact way misalignment is allowed to resolve.

From this point forward, the world is no longer explained as something that is pushed into motion. It is understood as something that moves because it must.

Chapter 4 — Resolution Is Becoming

When Change Does Not Feel Chosen

Most changes in the world do not feel chosen. They feel inevitable.

A crack spreads through a surface without hesitation. A liquid finds its level without deliberation. A hot object cools, a stretched body relaxes, a crowded system disperses. None of these events give the impression of decision. They unfold as if something simply had to give.

And yet, the language we use to explain them is full of choice.

We say a system "moves toward equilibrium," "seeks stability," or "prefers a lower energy state." These phrases are convenient, but they quietly insert intention where none exists. They suggest that nature evaluates options, selects outcomes, or aims for goals. In doing so, they obscure what is actually happening.

Nature does not choose. Nature resolves.

This difference matters more than it first appears. Choice implies alternatives. Resolution implies necessity. When a system changes, it is not weighing possibilities; it is relieving incompatibility. The path taken is not selected—it is the only one left open by the system's structure and constraints.

This is why change often feels impersonal yet precise. A system does not ask permission to reorganize, nor does it wait for observation. It unfolds whether or not anyone is watching, and it does so with a consistency that defies interpretation as decision-making. The outcome may surprise us, but it never surprises the system itself.

We tend to project agency onto these processes because we experience choice internally. We decide, hesitate, commit. But this internal experience does not scale outward. The mistake begins when we export the language of mind into the fabric of nature, mistaking *sequence* for *selection*.

The moment this projection is removed, something becomes clear: becoming is not an act. It is a process.

Resolution is the only thing that ever happens. Motion, flow, decay, growth, and formation are not separate phenomena; they are different faces of the same unfolding. They differ not in intention, but in how misalignment is allowed to resolve.

This chapter asks the reader to notice this inevitability—not as fatalism, but as clarity. Once change is understood as resolution rather than choice, the world stops appearing driven by hidden intentions and starts appearing governed by simple necessity.

Nothing is deciding. Nothing is aiming. Nothing is observing itself into existence.

Something is resolving.

And where resolution proceeds, becoming follows.

4.1 — Manifestation Begins at Resolution, Not Observation

One of the most persistent confusions in modern explanation is the belief that things somehow *come into being* when they are observed. This idea has gained strength not because it reflects reality, but because it offers psychological comfort. It places the observer at the center, turning uncertainty into participation and mystery into importance.

But the world does not wait to be seen.

Mountains erode whether or not anyone measures them. Stars collapse in silence. Chemical reactions complete themselves in sealed containers. Bubbles form, stretch, thin, and burst without witnesses. In every case, manifestation occurs not at the moment of observation, but at the moment resolution becomes unavoidable.

Observation records. Resolution creates.

This distinction restores causality to where it belongs. A system does not exist in a suspended state until an observer arrives. It exists in whatever condition its internal alignment allows. When misalignment appears, resolution begins immediately, regardless of awareness. Form emerges because resolution is underway, not because it has been noticed.

The confusion arises because observation and manifestation often coincide in human experience. We become aware of something at the same moment it becomes perceptible. But coincidence is not causation. The observer enters the story late—at the point where resolution has already produced detectable structure.

This is why observation feels powerful but is not generative. Seeing a falling object does not make it fall. Measuring a temperature gradient does not create heat flow. Watching a system reorganize does not initiate the process. These events were already unfolding. Observation merely aligns the observer with an ongoing resolution.

Understanding manifestation this way removes the need for special roles, collapses, or participatory mechanisms. Reality does not flicker into existence upon inspection. It unfolds continuously as long as misalignment demands it. The observer is simply another system that becomes aligned with that unfolding.

This also explains why manifestation feels objective rather than personal. Different observers see the same event because resolution is not tailored to perception. It is governed by structure, constraints, and delta—not by awareness.

When manifestation is anchored to resolution instead of observation, several long-standing confusions dissolve at once. The world regains independence. Change regains necessity. The observer returns to scale—not irrelevant, but not privileged.

Manifestation is not an act of seeing. It is the visible face of resolving incompatibility.

Once this is clear, the story of becoming can proceed without invoking mystery. The world does not need witnesses to unfold. It needs only one thing: something that cannot remain as it is.

And when that condition exists, manifestation follows—quietly, precisely, and inevitably.

4.2 — Why Systems "Decide" Nothing

It is tempting to speak of systems as if they make choices. We say a fluid *chooses* a path, a particle *decides* where to go, a system *selects* a stable state. This language feels natural because outcomes often look selective. One path is taken, another is not. One structure forms, another fails to appear. To the observer, this resembles decision-making.

But resemblance is not identity.

Decision implies alternatives evaluated from within. It implies preference, hesitation, intent. Physical systems possess none of these. They do not stand at crossroads weighing options. They encounter constraints, and within those constraints, only certain resolutions are possible. What looks like choice is simply necessity expressing itself.

A system never asks, "What should I do next?" It encounters the question, "What can no longer remain unresolved?"

The path that follows is not chosen; it is the only path that remains open given the system's structure, boundaries, and current misalignment. If multiple pathways exist, resolution distributes along them according to capacity—not preference. If only one pathway exists, resolution takes it. If no pathway exists, tension accumulates until one is created or the system breaks.

This is why outcomes feel decisive without being deliberate.

When a fluid flows around an obstacle, it does not decide to go left or right. It resolves pressure gradients through whatever geometry allows. When a crack propagates, it does not choose its direction; it follows stress concentration. When a chemical reaction proceeds, it does not select products; it follows compatibility of bonds and energy redistribution.

The illusion of decision arises because we observe final forms, not resolution processes. We see what happened and imagine alternatives that never truly existed under the given conditions. From the system's perspective, there were no real options—only allowable resolutions.

This distinction matters because decision-language quietly smuggles agency into explanations where none is needed. Once agency is assumed, questions multiply unnecessarily: *Why this outcome and not that one? What guided the choice? What principle selected the path?* These questions dissolve when choice is removed from the picture.

Nature does not select outcomes. Nature exhausts incompatibilities.

What remains after resolution is not the result of a decision, but the result of what could be resolved given the constraints. No preference is involved. No intelligence is required. Only structure and misalignment.

This also explains why similar systems under similar conditions produce similar outcomes reliably. If decision were involved, variation would be fundamental. Instead, variation appears only when conditions differ. Repeat the misalignment, repeat the resolution.

By removing decision from systems, we do not make the world mechanical or dull. We make it precise. Becoming is no longer mysterious or anthropomorphic. It is the exact unfolding of what cannot remain unresolved.

Systems do not decide to change. They change because they must.

And once necessity replaces choice, the direction of becoming becomes clear—because resolution has only one direction it can ever take.

4.3 — Resolution as the Only Direction Nature Takes

Once misalignment appears, nature has no freedom in the matter. It does not pause, reverse course, or explore alternatives. Resolution proceeds in one direction only—toward compatibility. This is not a tendency, not a preference, and not a law imposed from outside. It is a consequence of what misalignment means.

Misalignment cannot sustain itself indefinitely. It either resolves or accumulates until resolution becomes unavoidable.

This is why becoming feels irreversible. A system can return to a similar state, but it cannot undo the act of resolution itself. Once incompatibility has been relieved through reorganization, the path taken becomes part of history. The system may cycle, oscillate, or fluctuate, but each resolution leaves an imprint. Becoming does not rewind; it unfolds.

This irreversibility does not come from time pushing events forward. It comes from resolution exhausting misalignment. Once a particular configuration has been resolved, that exact incompatibility no longer exists in the same way. The system has changed its internal landscape. Going "back" would require recreating the same misalignment, which is not reversal but a new event.

This is why nature never resolves away from alignment. There is no mechanism that increases incompatibility for its own sake. What we sometimes interpret as growing disorder or increasing chaos is not nature choosing misalignment—it is delta exceeding resolution capacity. The direction remains the same; the system is simply unable to keep pace.

Even processes that appear destructive follow this rule. Decay, breakdown, erosion, and collapse are not movements toward disorder as a goal. They are resolutions of misalignment at scales where coherence can no longer be maintained. When a structure breaks, it is not choosing destruction; it is relinquishing a configuration it can no longer support.

This understanding unifies phenomena that are often treated separately:

- Dissipation is resolution of excess delta into broadly distributed alignment.
- Relaxation is resolution of internal strain into compatibility.
- Flow is resolution of gradients through accessible pathways.
- Equilibration is resolution reaching completion.

In every case, the direction is the same. The system moves from incompatibility toward compatibility—not because it wants to, but because there is nowhere else to go.

Importantly, this direction does not imply a final goal. Resolution does not aim for perfection or permanence. It simply proceeds until the current misalignment is exhausted. New deltas may appear, new resolutions may begin, but each act of becoming is driven by the same necessity.

Nature has no reverse gear for resolution. It does not oscillate between wanting and not wanting. It does not wander.

It resolves.

And because resolution has direction, becoming acquires direction without intention. What we call "the arrow of time" is nothing more than this fact made visible: misalignment resolves, and resolution does not undo itself.

Once this is understood, irreversibility stops being mysterious. It is not imposed by entropy, enforced by law, or carried by time. It is simply the unavoidable consequence of incompatibility being relieved.

Resolution is not something nature sometimes does. It is the only thing nature ever does when change occurs.

Everything else—motion, flow, decay, formation—is just how that one direction appears under different conditions.

4.4 — Time as the Measure of Resolution

Time is often treated as a backdrop—an invisible stage on which events unfold. We imagine it flowing uniformly, carrying motion along with it, governing change from the outside. But when examined closely, this picture collapses. Time does not drive resolution. Resolution generates time.

Where nothing resolves, time has no role to play.

A perfectly aligned system does not experience duration internally. Nothing progresses because nothing needs to. There is no sequence to mark, no before and after to distinguish. Stillness is not slow time; it is the absence of temporal experience. Time appears only when misalignment introduces necessity and resolution begins.

This is why time feels inseparable from change. It is not a separate dimension through which change moves; it is the record of change occurring. Each act of resolution creates a next state, and the ordering of these states is what we call time. Without resolution, there is no ordering. Without ordering, time has nothing to measure.

Seen this way, time is not universal in experience, even if clocks are universal in design.

Different systems resolve delta at different rates. A fast-resolving system compresses sequence, making time feel dense and urgent. A slow-resolving system stretches sequence, making time feel elongated and calm. Where resolution stalls, time feels suspended. Where resolution accelerates, time feels scarce. These are not psychological illusions layered on top of a fixed temporal flow; they are direct consequences of resolution rate.

This reframing also explains why time cannot be absolute. If time were fundamental, it would flow identically everywhere regardless of what was happening. But experience and measurement both contradict this. Time behaves differently across systems because resolution behaves differently across systems. Clocks do not measure time itself; they measure regular, repeatable resolution processes and use them as references.

A clock is not a window into time. It is a standardized resolver.

Once this is understood, the mystery around time's direction dissolves. Time does not move forward because it must. It moves forward because resolution does not reverse. Each resolved misalignment leaves behind a configuration that is no longer the same. Sequence accumulates naturally. The arrow of time is not imposed—it is inherited from the irreversibility of resolution.

Importantly, this view does not reduce time to perception. Time is not created by observers. Observers merely participate in resolution and experience its sequence from within. A process unfolding without witnesses still generates time for itself. Observation does not start the clock; resolution does.

Time, then, is neither a substance nor a container. It is a measure—an accounting of how incompatibility is relieved across a system. Where resolution is rich, time is vivid. Where resolution is sparse, time is thin. Where resolution completes, time falls silent.

This brings the chapter's core idea into full focus.

Becoming is resolution in motion. And time is how that motion is counted.

Nothing pushes time forward. Something resolves—and time appears as its trace.

4.5 — Becoming Without a Doer

Once resolution is recognized as the engine of change, a final illusion falls away: the need for a doer. We instinctively look for an agent behind every action—something that starts, steers, or sustains becoming. This habit comes from human experience, where intention precedes effort. But nature does not share this structure.

Becoming does not require a subject.

When misalignment appears, resolution proceeds without instruction. No entity initiates it, no intelligence supervises it, and no will sustains it. The process is automatic not because it is mechanical, but because necessity is sufficient. What cannot remain unresolved begins to reorganize the moment pathways exist.

This is why explanations that invoke hidden drivers—self-organization, emergent agency, guiding principles—feel unsatisfying. They replace one mystery with another. If order appears, we imagine something must be ordering it. If structure persists, we assume something must be maintaining it. In reality, persistence requires no maintenance as long as alignment holds. Order is not held together; it is what remains when resolution has finished its work.

The idea of a doer sneaks in through language. Verbs demand subjects. We say "the system relaxes," "the fluid flows," "the structure adapts." Grammar implies agency where none exists. What is actually happening is simpler: resolution unfolds according to constraints. The system is not acting; it is being acted *through* by necessity.

This perspective removes a significant burden from explanation. There is no need to account for intention in matter, no need to smuggle intelligence into particles, no need to elevate observation into creation. Becoming is not commanded. It is permitted and compelled by conditions.

Importantly, removing the doer does not make reality cold or empty. It makes it reliable. Processes unfold consistently because nothing interferes with necessity. Where constraints are the same, outcomes repeat. Where constraints differ, outcomes diverge. Variability arises from structure, not from choice.

This also clarifies the role of observers and living systems. Organisms do not create becoming; they are configurations through which resolution passes. What we experience as choice or agency at higher levels emerges from internal deltas resolving across complex structures. Even here, becoming precedes decision. Choice is a late, local phenomenon—not the source of change.

Nature does not need a hand to push it forward. It does not need a mind to tell it what to do. It does not need a witness to make it real.

Becoming happens because misalignment exists and resolution is unavoidable.

Once the doer is removed, becoming appears exactly as it is: a continuous, impersonal, and precise unfolding of compatibility. Nothing directs it. Nothing resists it. Nothing stands outside it.

What follows next is not agency, but pattern. Not intention, but regime. Resolution always proceeds—but how it proceeds depends on the size of delta and the capacity of the system to resolve it. That distinction opens the door to coherence, turbulence, and collapse.

With becoming clarified, the next chapter asks a sharper question: *When does resolution produce structure—and when does it destroy it?*

From Becoming to Regimes

This chapter has removed the last layer of unnecessary agency from our understanding of change. Becoming no longer needs intention, instruction, or observation to occur. It requires only one thing: misalignment that cannot be ignored. From that moment onward, resolution unfolds with exactness—neither choosing nor hesitating, neither accelerating nor slowing without cause.

What has emerged is a simple, rigorous picture.

Manifestation begins when resolution begins. Systems do not decide; they respond to necessity. Resolution moves in one direction only—toward compatibility. Time is not a container, but the measure of how resolution proceeds. Becoming needs no doer, because necessity is enough.

At this point, one might be tempted to conclude that all resolution looks the same—that once misalignment appears, nature always responds smoothly and predictably. But experience contradicts this. Sometimes resolution produces elegant flow and stable structure. Sometimes it fragments into turbulence, noise, or collapse. Sometimes it barely proceeds at all, breaking into jumps and fluctuations.

Resolution is inevitable. Its outcome is not uniform.

What determines the difference is not energy, not force, and not intention. It is the relationship between delta and the system's capacity to resolve it. When that relationship is compatible, coherence appears. When it is strained, instability follows. When it is overwhelmed or starved, order fails in distinct ways.

The next chapter introduces this decisive distinction by mapping the regimes of delta. It shows that chaos, coherence, and fragmentation are not separate phenomena, but expressions of the same process operating under different conditions. This classification becomes a universal reference point—not just for physics, but for any system that changes, adapts, or fails.

Becoming has now been understood as resolution in motion. What remains is to understand how resolution behaves.

Nature does not act. Nature resolves.

With that clarity in place, we are ready to see why the same world can appear calm, turbulent, or fragile—and why none of those states are accidental.

Chapter 5 — The Three Regimes of Delta

When the Same World Behaves Radically Differently

There are moments when the world feels smooth and cooperative. Things flow. Effort produces results. Systems respond proportionally. At other times, the same world feels unstable. Small actions trigger disproportionate reactions. Structures collapse. Processes spiral into chaos. And sometimes, nothing seems to move at all—only jitter, noise, and abrupt jumps without continuity.

These differences are so familiar that we rarely question them. We accept turbulence, stagnation, and harmony as separate behaviors, governed by different rules. When things go wrong, we add more force. When that fails, we invoke randomness. When stability appears, we treat it as an exception.

But the world has not changed its rules.

What has changed is the relationship between misalignment and the system's capacity to resolve it.

This chapter begins with a simple observation: the same system, under the same laws, can behave in radically different ways without invoking new causes. Flow can turn into turbulence. Order can dissolve into breakdown. Stability can fragment into noise. These transitions do not require new forces or hidden variables. They arise naturally when delta enters a different regime.

We are accustomed to thinking in terms of more or less—more energy, more pressure, more effort. But resolution does not respond linearly to magnitude. It responds to fit. When misalignment exceeds capacity, coherence fails. When misalignment is too weak, continuity cannot be sustained. Only within a narrow band does resolution proceed smoothly.

This explains a deep intuition many people already have: pushing harder often makes things worse, while pushing too gently does nothing at all. Whether in fluids, materials, learning, or social systems, effectiveness depends less on intensity than on *timing and scale*. The world rewards alignment, not force.

What we often call chaos is not disorder for its own sake. It is resolution overwhelmed. What we call noise is not randomness. It is resolution starved. And what we call order is not imposed structure, but resolution in balance.

This chapter does not introduce new behavior. It organizes familiar behavior under a single, unifying lens. By doing so, it removes the false opposition between order and chaos and reveals them as neighboring regimes of the same process.

Once these regimes are seen clearly, instability stops feeling mysterious, coherence stops feeling miraculous, and failure stops feeling personal. They are no longer moral or metaphysical conditions. They are physical outcomes of how delta is applied.

The same world behaves differently not because it changes its nature, but because delta crosses a boundary.

Understanding those boundaries is the key to understanding everything that follows.

5.1 — Delta Is Not Good or Bad: It Is Conditional

Delta is often treated as a problem to be eliminated or a stimulus to be maximized. When things fail, we assume there was “too much” disruption. When nothing happens, we assume there was “too little” effort. Both instincts miss the point. Delta is neither beneficial nor harmful by itself. Delta is necessary, but it must be *matched* to the system that receives it.

Without delta, nothing changes. With delta, change becomes unavoidable.

This alone makes delta indispensable. Every movement, every transformation, every act of becoming begins only because some form of misalignment appears. A perfectly aligned system is complete; it has no reason to move. Delta is the invitation to reconfigure. But invitation does not guarantee success.

What matters is not the presence of delta, but its relationship to resolution capacity.

A system can absorb only so much misalignment at a time. That capacity depends on structure, coupling, pathways, and internal coherence. When delta enters within that capacity, resolution proceeds smoothly. When delta exceeds it, coherence breaks. When delta is too weak, resolution cannot sustain continuity. In all three cases, the same principle is operating—only the *regime* has changed.

This reframes a common misunderstanding: instability is not evidence that delta is "bad," and stillness is not evidence that delta is "missing." Instability arises when delta overwhelms alignment. Stillness arises when delta is absent or too weak to matter. Both are natural outcomes, not failures of the system or errors in nature.

Delta does not scale morally. It scales structurally.

This is why increasing delta does not reliably improve outcomes. More pressure does not guarantee better flow. More effort does not guarantee learning. More force does not guarantee progress. Past a threshold, additional delta degrades resolution instead of enhancing it.

Likewise, reducing delta too far does not preserve stability. Systems with insufficient delta drift into fragmentation—small, discontinuous adjustments without coherence. Noise replaces flow. Jumps replace continuity. Activity appears, but nothing accumulates.

Understanding delta as conditional rather than evaluative dissolves a false dichotomy. Delta is not something to fear or worship. It is something to respect. It must be applied in proportion to what the system can resolve.

This insight sets the stage for the three regimes that follow. Chaos, noise, and coherence are not separate behaviors governed by different rules. They are expressions of the same process operating under different delta–capacity relationships.

Delta is always doing the same thing: inviting resolution.

What changes is whether the system can accept that invitation.

5.2 — Regime One: Large Delta — Chaos, Turbulence, Breakdown

When delta exceeds a system's capacity to resolve it, coherence does not fail politely. It fragments.

Large delta does not produce stronger order; it overloads resolution pathways. The system is confronted with more misalignment than it can reorganize coherently within its structure and timing. Resolution still occurs—nature never stops resolving—but it does so without coordination. The result is what we recognize as chaos, turbulence, or breakdown.

This is a crucial correction: chaos is not randomness. Chaos is resolution under excess.

In this regime, misalignment arrives faster or larger than the system can distribute. Instead of flowing smoothly, resolution breaks into competing local adjustments. Different parts of the system attempt to resolve delta independently, without shared coherence. The global structure cannot hold, so it fractures into transient patterns that form and dissolve rapidly.

This is why turbulence looks busy but ineffective. Enormous activity occurs, yet little stabilizes. Energy is abundant, motion is intense, but alignment is scarce. Resolution is happening everywhere—and succeeding nowhere.

Breakdown is simply the extreme form of this regime. When excess delta overwhelms not just pathways but structure itself, the system relinquishes its configuration entirely. Materials crack. Organizations collapse. Minds burn out. Ecosystems fail. None of this happens because resolution stopped. It happens because resolution had no coherent way forward.

Importantly, large delta does not always announce itself loudly. It can arrive subtly through rate rather than magnitude. A moderate misalignment introduced too quickly can overwhelm a system just as effectively as a large one applied slowly. This is why systems often fail "suddenly," even though nothing dramatic appears to have changed. The threshold was crossed in timing, not quantity.

This regime explains a common but dangerous mistake: trying to fix instability by applying more force. When coherence is already lost, adding delta only deepens fragmentation. More pressure increases turbulence. More control accelerates collapse. More urgency destroys remaining alignment.

Chaos is not nature misbehaving. It is nature trying to resolve more than the system can handle.

Once seen this way, chaos loses its mystique. It is not an enemy, not a curse, and not an exception. It is a predictable outcome when delta outruns resolution capacity. Remove the overload—by reducing magnitude, slowing rate, or increasing capacity—and coherence can reappear just as naturally.

Large delta does not create disorder by intention. It simply leaves resolution with no coordinated path.

Understanding this regime is the first step in learning why stability cannot be forced—and why coherence is always a matter of fit, never of dominance.

5.3 — Regime Two: Too Small Delta — Jumps, Noise, Fluctuations

If excess delta overwhelms a system, insufficient delta starves it. In this regime, misalignment exists, but not strongly enough—or not continuously enough—to sustain smooth resolution. The system does not settle into silence, yet it cannot maintain flow. What emerges instead is fragmented activity: small jumps, jitter, noise, and intermittent fluctuations.

This regime is often mistaken for randomness.

In reality, it is resolution attempting to proceed without enough drive. Delta appears in tiny bursts or at irregular intervals. Each burst triggers a small adjustment, but before coherence can build, the misalignment fades or shifts. Resolution keeps starting—and keeps stopping. Continuity never establishes itself.

This is why behavior here looks jumpy rather than fluid.

In physical systems, this appears as stick–slip motion, micro-vibrations, thermal noise, or discrete transitions instead of smooth change. In fluids near rest, motion breaks into eddies that never organize into flow. In materials, deformation occurs in tiny slips rather than continuous strain.

In cognitive systems, this regime shows up as distraction, indecision, and scattered thought. Effort is present, but not enough to carry understanding forward. Ideas appear briefly, then vanish. Nothing integrates. In social systems, it appears as constant chatter without movement—activity without progress.

What defines this regime is not inactivity, but incoherence.

The system is trying to resolve misalignment, but delta is too weak to overcome internal friction, thresholds, or structural inertia.

Each attempt at resolution dies before it can propagate. The result is a landscape filled with local, short-lived adjustments that never connect.

This explains why noise often increases near stillness. When a system is almost—but not quite—aligned, tiny deltas flicker across it. These flickers are real, but they cannot accumulate into motion. Observed from the outside, the system looks unpredictable. From within, it is simply under-driven.

Importantly, adding a little more delta does not immediately produce coherence. There is a threshold that must be crossed. Below it, resolution remains discontinuous. This is why encouragement, nudges, or minor interventions often fail to produce change. The system is not resisting; it is unable to sustain resolution.

Too-small delta is not harmless. Prolonged exposure to this regime erodes structure. Constant micro-adjustments fatigue systems without building anything lasting. Noise consumes attention. Fluctuations drain capacity. Over time, the system becomes less able to respond even when stronger delta finally arrives.

This regime teaches an essential lesson: change requires enough misalignment to matter.

Silence comes from completion. Coherence comes from matched delta. But noise comes from delta that is present yet insufficient.

Understanding this regime prevents a common error: mistaking activity for progress. Not all motion builds. Not all effort flows. Without adequate delta, resolution cannot commit to a path—and without commitment, becoming remains fragmented.

Too little delta does not preserve stability. It dissolves it quietly, one fluctuation at a time.

5.4 — Regime Three: Right Delta — Coherence, Flow, Structure

Between overload and starvation lies a narrow, decisive band. Here, delta is neither excessive nor insufficient. It arrives at a rate and scale the system can absorb, distribute, and resolve without losing coordination. In this regime, resolution becomes coherent—and coherence changes everything.

This is where flow appears.

Right delta does not force change. It allows change to propagate. Misalignment is strong enough to overcome inertia and thresholds, yet gentle enough to remain shared across the system. Resolution does not fragment into competing local fixes, nor does it stall into flicker. It moves as a coordinated wave, aligning parts as it proceeds.

What emerges feels intelligent, even though no intelligence is directing it.

In physical systems, this regime produces laminar flow, stable oscillations, resilient structures, and smooth transitions. In fluids, motion organizes instead of breaking apart. In materials, deformation distributes evenly rather than cracking. In thermal systems, gradients relax without turbulence.

In biological systems, right delta sustains life. Metabolism proceeds steadily. Growth integrates. Adaptation refines rather than destabilizes. The organism neither burns out nor stagnates because internal misalignment is continuously resolved within capacity.

In cognitive systems, this is the state of clarity and learning. Challenge is present but not overwhelming. Attention holds. Understanding accumulates. Thought flows without force. The mind is neither anxious nor bored—it is aligned with the work of resolution.

In social systems, this regime produces trust, cooperation, and durable institutions. Change occurs without collapse. Reform proceeds without revolt. Structure evolves without losing coherence. Progress feels organic rather than imposed.

The defining feature of this regime is continuity.

Resolution in right delta does not need to restart itself. Each step prepares the ground for the next. The system carries memory forward as structure. Patterns stabilize. What was once dynamic becomes reliable. This is how structure forms—not by freezing motion, but by remembering how resolution flowed.

Coherence is fragile because the window is narrow. A little more delta, and the system tips into chaos. A little less, and it falls into noise. But coherence is also persistent, because once established, it expands capacity. Structures built in this regime increase the system's ability to handle future misalignment. The system learns, thickens, stabilizes.

This is why coherence feels alive. It is resolution that has found its pace.

Importantly, coherence is not equilibrium. It is active alignment. Resolution is ongoing, but it never accumulates strain. Time is present, but not oppressive. Motion exists, but without violence. Change occurs, but without rupture.

Right delta does not eliminate instability. It integrates it.

Understanding this regime corrects a common mistake: believing that stability requires eliminating misalignment. In reality, stability requires the right amount of misalignment, continuously resolved. A perfectly aligned system is silent. A coherent system is alive.

This regime is not rare in nature. It is what nature prefers whenever conditions allow. Chaos and noise are not competitors; they are failures of fit. Coherence is what appears when delta and capacity meet precisely.

Right delta is not balance by cancellation. It is balance by flow.

And where flow persists, structure follows—not as an imposition, but as a memory of successful resolution.

5.5 — Transitions Between Regimes

The most striking feature of delta regimes is not their existence, but how abruptly systems move between them. Change does not usually slide smoothly from coherence into chaos, or from noise into flow. Instead, systems hold their behavior for long periods and then shift suddenly—as if a line has been crossed.

This is because regimes are separated by thresholds, not gradients.

A system can tolerate increasing misalignment without visible change until a critical limit is reached. Below that limit, resolution remains coordinated. Above it, coherence collapses. The same is true in reverse: reducing delta does not immediately restore order once chaos has set in. The system must regain enough alignment capacity before coherent resolution becomes possible again.

These thresholds are not arbitrary. They are determined by structure, coupling, timing, and internal bandwidth. What changes at the threshold is not the nature of resolution, but its ability to remain shared across the system.

This explains why transitions often feel dramatic.

A fluid remains laminar until turbulence appears suddenly. A structure holds load until it snaps. A person functions until burnout arrives. An institution appears stable until it collapses almost overnight.

In each case, delta did not suddenly appear. It accumulated or accelerated quietly. What changed was the system's ability to resolve it coherently. Once that ability was exceeded, the regime shifted.

Importantly, transitions are asymmetric.

It is often easier to fall out of coherence than to return to it. Chaos degrades structure. Noise erodes continuity. When resolution fragments, the pathways that once supported smooth flow are damaged or lost. Restoring coherence requires rebuilding capacity, not merely reducing delta. This is why recovery takes longer than collapse.

Transitions also depend on rate, not just magnitude. A system may handle a large delta if it arrives slowly, yet fail under a smaller delta applied too quickly. Timing determines whether resolution can distribute misalignment or is forced into local, uncoordinated responses. This is why sudden shocks are so destructive—and why gradual change often succeeds where force fails.

Understanding transitions dissolves another common misconception: that systems fail because something "went wrong." Most failures are not errors. They are regime crossings. The system did exactly what it could until it no longer could.

Likewise, recovery is not reversal. A system does not go back to its previous state. It must establish a new coherent regime under new conditions. Even when coherence returns, it is not the same coherence as before. Resolution has already left its mark.

Transitions teach a final, crucial lesson: there is no single optimal behavior.

What is coherent in one regime is disastrous in another. Strategies that work near flow fail near chaos. Techniques that stabilize noise suffocate coherence. This is why one-size-fits-all explanations collapse under real conditions.

Regimes do not change because nature is fickle. They change because capacity is crossed.

Recognizing regime transitions allows us to stop fighting symptoms and start respecting thresholds. It shifts attention from control to fit, from force to timing, from intensity to readiness.

With this understanding, chaos is no longer mysterious, noise is no longer trivial, and coherence is no longer accidental. They are simply the natural faces of resolution operating under different conditions—and moving between them when limits are reached.

The next step is unavoidable: if transitions depend on capacity over time, then time itself must have structure. That structure is not universal. It belongs to the system.

That insight leads directly to Alignment Threshold Time (ATT).

5.6 — Scale Independence of Delta Regimes

One of the strongest tests of any foundational idea is whether it fractures when scale changes. Many explanations work beautifully at one level—particles, fluids, organisms, societies—only to fail when stretched beyond their comfort zone. Delta regimes do not suffer from this weakness. They repeat with unsettling consistency, regardless of scale.

This is not analogy. It is structural recurrence.

At the smallest physical scales, delta regimes appear in transitions between stable states, fluctuations near thresholds, and breakdown under excessive perturbation. At intermediate scales, they govern fluid flow, material deformation, and thermal transport. At large scales, they shape ecosystems, economies, institutions, and civilizations. The language changes, but the behavior does not.

Large delta overwhelms resolution everywhere.

- In physics, it produces turbulence, shock, fracture.
- In biology, it produces stress responses, failure, death.
- In cognition, it produces panic, burnout, collapse of clarity.
- In societies, it produces unrest, violence, systemic breakdown.

Too-small delta starves resolution everywhere.

- In physics, it produces noise, stick–slip motion, thermal

jitter.

- In biology, it produces stagnation, decay, loss of adaptation.
- In cognition, it produces distraction, indecision, rumination.
- In societies, it produces inertia, performative activity, hollow motion.

Right delta enables coherence everywhere.

- In physics, it produces flow, structure, stable patterns.
- In biology, it produces metabolism, growth, resilience.
- In cognition, it produces learning, insight, sustained attention.
- In societies, it produces cooperation, trust, durable progress.

The repetition is too precise to dismiss as metaphor.

What varies across scale is not the role of delta, but the form of capacity. A fluid resolves misalignment through viscosity and geometry. A mind resolves it through attention and integration. A society resolves it through institutions and norms. Different mechanisms—same condition, same regimes.

This is why explanations that rely on domain-specific causes eventually fail. They describe *how* resolution happens locally, but they cannot explain *why* behavior collapses into the same patterns everywhere. Delta regimes provide that missing unifier without flattening complexity.

Scale independence also explains why lessons transfer across domains so naturally. Why strategies learned in physics reappear in psychology. Why organizational failure mirrors material failure. Why personal burnout feels structurally identical to mechanical overload. These are not poetic parallels; they are regime crossings governed by the same logic.

Importantly, this universality does not erase specificity. Each system has its own thresholds, capacities, and pathways. Right delta for one system may be chaos for another. What is coherent at one scale may be destructive at another. Delta regimes do not standardize behavior; they contextualize it.

Recognizing scale independence corrects two common errors at once.

First, it prevents over-specialization—believing that every domain needs entirely new principles. Second, it prevents over-generalization—believing that one rule applies identically everywhere.

Delta regimes occupy the middle ground: one condition, many expressions.

This is why this chapter functions as a reference point. Once the regimes are understood, they can be recognized instantly in any system. The question shifts from "What is happening?" to "Which regime are we in—and why?"

At that moment, explanation becomes practical.

Scale does not change the logic of becoming. It only changes the language we use to describe it.

With regimes now mapped and their universality established, one final refinement is needed. Regimes explain *what kind* of behavior appears. They do not yet explain *when* a system tips from one to another.

That timing belongs to the system itself.

And that is where Alignment Threshold Time enters.

5.7 — Why This Is the Missing Classification

For a long time, instability, noise, and order have been treated as fundamentally different phenomena. Chaos was given its own mathematics. Noise was treated as error or randomness. Order was celebrated as structure or equilibrium. Each received separate explanations, separate tools, and separate vocabularies. What was missing was not detail—but classification at the right level.

Delta regimes provide that missing layer.

They do not replace existing theories; they sit beneath them. They explain why wildly different systems—governed by different laws, equations, and constraints—fall into the same behavioral patterns. Turbulence in fluids, burnout in people, market crashes, and material fracture were never separate problems. They were the same condition expressing itself through different structures.

Traditional frameworks describe behavior after the fact. Delta regimes classify conditions before outcomes diverge.

This is why chaos and order have always felt uncomfortably related. They were never opposites. They are neighbors. They sit on either side of coherence, separated only by mismatch between delta and resolution capacity. What changes is not the rules of nature, but whether those rules can operate coherently under current conditions.

Without this classification, explanation fragments. We invent special causes for breakdown, special models for noise, and special narratives for success. Failure becomes moralized. Stability becomes romanticized. Chaos becomes mystified. Delta regimes remove all three distortions.

They show that:

- Chaos is not disorder, but resolution overwhelmed
- Noise is not randomness, but resolution starved
- Order is not imposed, but resolution aligned

This unification is why the framework feels immediately applicable. It does not ask the reader to learn new behaviors—only to recognize familiar ones under a clearer lens. Once seen, the regimes are difficult to unsee. Every instability, every stagnation, every flow announces its regime plainly.

Most importantly, this classification restores predictive intuition without pretending to offer control. It does not promise that coherence can always be created, or chaos always avoided. It shows instead *why* certain attempts fail and *why* others succeed—often for reasons that intensity, intelligence, or intention alone cannot overcome.

By naming regimes rather than outcomes, the framework shifts focus:

- from results to conditions
- from force to fit
- from effort to capacity

This is why the classification is missing but necessary. Without it, explanations remain reactive. With it, understanding becomes anticipatory. One can see instability approaching before it erupts, recognize stagnation before it drains capacity, and protect coherence before it collapses.

Delta regimes do not simplify the world by denying complexity. They simplify it by placing complexity where it belongs.

This chapter completes a structural triad:

- Previous chapter explained *what* becoming is (resolution).
- This chapter explains *how* becoming behaves (regimes).
- Next chapter will explain *when* becoming fails or succeeds (time and thresholds).

With regimes now clearly defined, the final missing piece is timing. Not clock time, but system time—the bandwidth within which resolution can remain coherent.

That bandwidth has a name.

And it is the key to understanding why patience, urgency, collapse, and resilience are not moral qualities, but physical realities.

From Regimes to Thresholds

This chapter has completed a classification that was long overdue. Chaos, noise, and coherence are no longer treated as separate mysteries, moral failures, or special cases. They are revealed as natural regimes of the same process, distinguished only by how delta relates to a system's capacity to resolve it.

Nothing new had to be invented for this to become clear. Nothing familiar had to be discarded.

Once delta is placed correctly, behavior organizes itself.

Large delta overwhelms coordination and fractures resolution into chaos. Too-small delta starves continuity and fragments resolution into noise. Right delta allows resolution to propagate coherently, giving rise to flow and structure.

These are not outcomes chosen by nature. They are conditions crossed by systems.

At this point, one essential question remains unanswered. If regimes depend on capacity, then what defines that capacity? Why does the same delta feel overwhelming to one system and manageable to another? Why does timing matter more than magnitude? Why does slow change succeed where sudden change fails?

The answer cannot lie in delta alone.

What governs regime shifts is not just *how much* misalignment exists, but how quickly a system can align in response. Resolution has bandwidth. Systems do not fail because they lack energy or intelligence; they fail because resolution arrives faster than alignment can occur.

This brings time back into the story—but not as an abstract dimension or universal clock. Time returns as a system-specific constraint, inseparable from structure and capacity. The next chapter formalizes this constraint and gives it a name: Alignment Threshold Time (ATT).

Regimes tell us *what kind* of behavior appears. ATT tells us *when* behavior breaks, stabilizes, or flows.

With regimes now clearly mapped, the book moves to its final structural foundation. Not intention. Not force. But timing.

Because in the end, coherence is not about doing the right thing. It is about allowing resolution to happen at the rate the system can sustain.

Delta does not destroy order. Mismatch between delta and resolution capacity does.

Chapter 6 — Alignment Threshold Time (ATT)

Why Time Always Feels Like the Problem

When things fail, people rarely say they lacked energy or intent. They say something more telling: *it happened too fast*, or *it never got the time it needed.* Deadlines overwhelm, pressure accumulates, learning stalls, relationships fracture—not because nothing was done, but because timing collapsed.

Time is blamed constantly, yet almost never understood.

We speak of "running out of time," "buying time," "needing more time," as if time were a resource that could be stored, spent, or controlled. But two systems given the same duration do not behave the same way. One stabilizes. Another breaks. One integrates. Another fragments. The clock shows identical numbers; the outcomes diverge completely.

This alone should raise suspicion.

If time were absolute—flowing identically everywhere—then identical durations would produce comparable results. They do not. What matters is not how much time passes, but how much resolution a system can complete within that time.

This is why urgency destroys coherence while patience restores it. Not because patience is virtuous, but because it respects a physical constraint. Resolution cannot be rushed beyond a system's ability to align. When it is, fragmentation replaces flow. Structure collapses into noise or chaos, even when intentions are good and effort is high.

The discomfort we feel around time is not psychological. It is structural. We sense, often correctly, that something is being forced faster than it can settle. We feel misalignment growing, even when progress appears outwardly successful. Burnout, brittleness, and sudden failure are all signatures of this violation.

Yet most frameworks treat time as external and uniform. They optimize schedules, compress timelines, and celebrate speed without asking a more fundamental question: speed relative to what?

This chapter introduces Alignment Threshold Time (ATT) to answer that question. ATT is not about clocks or calendars. It is about resolution bandwidth—the maximum rate at which a system can absorb delta without losing coherence. Exceed it, and no amount of intelligence, control, or effort will prevent breakdown. Respect it, and systems stabilize, grow, and even expand their own capacity over time.

Seen this way, time stops being an enemy. It becomes a signal.

When things feel rushed, it is not because time is scarce. It is because alignment is being violated.

Understanding this difference changes everything. It turns patience from a moral demand into a physical strategy, and urgency from a virtue into a risk factor. Most importantly, it reveals why so many failures occur despite doing "everything right."

The problem was never time itself. The problem was forcing resolution faster than the system could align.

With that clarity, we can finally define what time really constrains—and why ignoring that constraint is the fastest way to collapse coherence.

6.1 — Why Time Is Not Absolute

The idea of time as something absolute is deeply ingrained. Clocks tick uniformly. Calendars advance predictably. Physics often treats time as a shared parameter against which all motion is measured.

This creates the impression that time flows independently of what is happening—that systems merely move *within* time.

But experience quietly contradicts this view.

Two systems subjected to the same duration do not undergo the same change. One stabilizes while another collapses. One learns while another fractures. One integrates while another becomes brittle. If time were truly absolute in its effect, this divergence would not be possible.

What remains constant is the clock—not the outcome.

This reveals the first crack in the notion of absolute time. Clocks measure repetition, not transformation. They count regular cycles—oscillations of atoms, vibrations of crystals, rotations of machinery. These cycles are useful reference processes, but they are not time itself. They do not govern change; they merely label sequences.

Actual change is governed by resolution, and resolution does not proceed at the same rate everywhere.

Time, as experienced by a system, is the ordering of its internal resolutions. Where resolution proceeds smoothly, time feels continuous and spacious. Where resolution struggles, time feels compressed or fractured. Where resolution stalls, time feels stagnant. These differences are not illusions layered on top of a universal flow—they are the direct expression of how resolution unfolds.

This explains why time "slows down" under stress and "disappears" during flow. It is not perception playing tricks. It is alignment changing pace. When resolution accelerates beyond capacity, sequence fragments. When resolution aligns perfectly, sequence feels effortless.

Absolute time would imply uniform causality: equal durations, equal capacity to change. Reality shows otherwise. Causality depends on structure, pathways, and bandwidth—not on the ticking of an external clock.

This also clarifies why time dilation appears in physics without invoking metaphysical paradoxes. Systems undergoing different resolution conditions will experience different ordering of events. The clock records consistency; the system experiences coherence or strain.

Time is not the river carrying events forward. It is the footprint left by resolution as it unfolds.

Once this is understood, the obsession with controlling time begins to look misplaced. You cannot manage time directly because time does not do anything by itself. What you can manage—if at all—is how resolution is allowed to proceed.

This reframing does not deny measurement or physics. It simply reassigns primacy. Time is no longer the driver of change; it is the record of change.

With this clarity in place, the next step becomes possible: defining the rate limit at which resolution can occur without losing coherence. That limit is not universal.

It belongs to the system.

And that limit is called Alignment Threshold Time.

6.2 — Alignment Threshold Time (ATT): Definition

Alignment Threshold Time (ATT) is the maximum rate at which a system can resolve delta while remaining coherent. It is not a duration measured on a clock, and it is not a subjective feeling of readiness. It is a structural bandwidth—a limit set by how a system is organized, coupled, and able to distribute misalignment internally.

Every system has one, whether recognized or not.

ATT answers a question that conventional models never ask: *How fast can resolution occur before alignment fails?* Below this threshold, resolution propagates smoothly. At the threshold, coherence is fragile but intact. Beyond it, resolution fragments—regimes shift, turbulence appears, noise increases, or breakdown follows.

This makes ATT fundamentally different from time as it is usually conceived. Time tells us *how long* something lasted. ATT tells us *how fast* change can safely happen. One is descriptive; the other is constraining.

ATT is not about patience as a virtue. It is about alignment as a physical necessity.

When delta arrives faster than ATT allows, the system is not choosing to fail. It is being asked to reorganize at a rate it cannot sustain. Resolution still occurs, but it loses coherence. Instead of coordinated alignment, the system responds through local fixes, shortcuts, or structural surrender.

This definition immediately clarifies several persistent confusions.

First, it explains why high-energy, high-intent interventions often fail. Effort increases delta, but effort does not expand ATT. Pushing harder does not widen resolution bandwidth. It often narrows it further by introducing stress, noise, and internal conflict.

Second, it explains why slow processes succeed where fast ones collapse. Slowness is not magical. It simply keeps resolution within ATT, allowing alignment to complete at each step before the next misalignment arrives.

Third, it explains why systems can tolerate large delta if it is spread over time, but fail under smaller delta delivered abruptly. ATT constrains *rate*, not magnitude. What matters is not how much change occurs, but whether alignment has time to settle before the next demand arrives.

ATT is emergent, not imposed. It arises from:

- the number and strength of internal pathways,
- the degree of coupling between components,

- the presence of memory or structure that carries resolution forward,
- and the system's current state of coherence.

A well-aligned system has a wider ATT. A degraded system has a narrow one. This is why stress compounds. Each ATT violation damages structure, reducing future capacity and making subsequent violations more likely.

Seen this way, ATT is the hidden governor behind stability and collapse. It is why schedules fail, why reforms backfire, why learning stalls, and why recovery takes time even after pressure is removed.

ATT does not ask whether change is desirable. It asks whether change is processable.

Once this limit is crossed, no amount of intelligence, morality, or control can restore coherence instantly. Alignment cannot be skipped. It must be allowed to complete at a pace the system can sustain.

With ATT defined, instability no longer looks mysterious. It becomes measurable—not in seconds or minutes, but in violations of resolution bandwidth.

The next section shows why this bandwidth is never universal, and why every system carries its own timing signature—whether it acknowledges it or not.

6.3 — ATT Is System-Specific

Alignment Threshold Time is not a constant that can be copied, optimized globally, or enforced uniformly. It belongs to the system itself. Two systems exposed to the same delta at the same clock speed can experience entirely different outcomes—not because time behaves differently, but because their capacities to align are different.

ATT is shaped by structure.

A system with many well-connected pathways can distribute misalignment smoothly. A system with rigid, narrow, or damaged pathways cannot. In the first case, resolution propagates; in the second, it bottlenecks. The clock does not change. The system does.

This is why ATT varies so widely across domains.

A fluid with low viscosity can absorb rapid gradients and still flow. A viscous fluid cannot. A trained mind can integrate complexity at a pace that overwhelms a novice. A mature organization can absorb reform that destabilizes a fragile one. The principle is the same: capacity determines timing.

ATT depends on several internal features:

- **Coupling density**: How well components communicate and share load
- **Pathway redundancy**: Whether resolution has multiple routes or only one
- **Memory and structure**: Whether past resolution stabilizes future alignment
- **Current coherence**: Whether the system is already strained or settled

These features are not static. ATT evolves.

When a system operates near its ATT without exceeding it, structure strengthens. Pathways thicken. Memory accumulates. The system becomes capable of resolving delta more quickly in the future. This is how learning, adaptation, and resilience develop.

When a system repeatedly violates ATT, the opposite occurs. Pathways degrade. Noise increases. Internal coupling weakens. Even modest delta begins to feel overwhelming. The system becomes brittle, reactive, and eventually unstable.

This explains a phenomenon often misunderstood as "loss of motivation" or "failure of will." What has actually occurred is ATT shrinkage. The system's bandwidth has been damaged by repeated over-speed resolution attempts.

ATT also explains why recovery cannot be rushed. Removing delta does not instantly restore capacity. Structure must rebuild. Pathways must re-form. Coherence must return. Until that happens, the system's ATT remains low, and even small demands can trigger instability.

Importantly, ATT is scale-dependent without changing principle. As systems grow more complex, their potential ATT can increase—but only if internal alignment grows with complexity. Otherwise, complexity narrows ATT instead of expanding it.

This is why scaling fails so often. What works for a small system breaks a larger one when alignment does not scale alongside delta.

ATT is not fairness. It is not preference. It is not negotiable.

It is a physical property of organization.

Recognizing that ATT is system-specific removes a dangerous assumption: that one timeline fits all. It replaces impatience with precision and urgency with respect for structure.

With this understanding, instability becomes predictable. The next section shows how most breakdowns are not caused by "too much change," but by a simple, often invisible mismatch: delta arriving faster than ATT allows.

6.4 — Delta–ATT Mismatch: The Root of Instability

Instability rarely begins with excess energy or lack of intent. It begins with a mismatch—delta arriving at a rate the system cannot align. When delta exceeds ATT, coherence does not degrade gradually; it fails structurally.

This is the core mechanism behind turbulence, noise, burnout, collapse, and brittleness across domains.

Delta–ATT mismatch means resolution is being demanded faster than alignment can complete. The system is not refusing to change; it is being forced to reorganize before its previous reorganization has settled. As a result, resolution fragments. Local fixes replace global coordination. Shortcuts replace structure. The system remains active but loses coherence.

Crucially, mismatch can occur in two directions.

When delta outruns ATT, instability appears as overload. Resolution breaks into competing local responses. Turbulence replaces flow. Stress accumulates. The system looks busy, even productive, but nothing stabilizes. This is the signature of chaos—not as disorder, but as over-speed resolution.

When delta lags far behind ATT, instability appears as under-drive. The system has capacity but no sustained misalignment to work through. Resolution flickers without commitment. Noise, jitter, and discontinuous jumps appear. The system looks active but directionless. This is not peace; it is resolution starvation.

In both cases, the system is unstable—not because delta is "wrong," but because timing is incompatible.

This reframes many familiar failures. Projects collapse not because goals were unrealistic, but because change was compressed beyond alignment bandwidth. Learning fails not because intelligence was lacking, but because complexity was introduced faster than integration could occur. Institutions fracture not because reform was unnecessary, but because reform outpaced coherence.

The most dangerous aspect of delta–ATT mismatch is that it often looks like progress—right until collapse. Early success masks accumulating strain. Metrics improve while alignment erodes. By the time symptoms are visible, structure has already been compromised.

This also explains why stabilization attempts frequently backfire. Applying more force increases delta without expanding ATT. Adding controls accelerates mismatch. Demanding urgency deepens fragmentation. The system appears resistant, but it is actually protecting itself from further damage.

Stability is not restored by reducing delta alone. It is restored by re-synchronizing delta with ATT. This can be done by slowing rate, staging change, increasing internal coupling, or allowing resolution to complete before introducing new demands. Without this resynchronization, instability persists regardless of intent.

Delta–ATT mismatch is not a mistake nature corrects gently. It is a boundary nature enforces strictly.

Once crossed, the system changes regime. Coherence cannot be commanded back into existence. It must be rebuilt at a pace the system can sustain.

Understanding this root cause removes blame from systems that "failed." They did not lack effort, intelligence, or will. They were subjected to incompatible timing.

With mismatch identified, the next question becomes practical and unavoidable: if timing governs stability, then qualities like patience and urgency must be reinterpreted—not as ethics, but as physics.

6.5 — Why Patience Is Physical, Not Moral

Patience is often framed as a personal virtue, a sign of discipline, humility, or emotional maturity. Impatience, in contrast, is treated as a flaw—something to be corrected through attitude or ethics. This framing misses the core issue. Patience is not primarily a moral quality. It is a physical requirement imposed by Alignment Threshold Time.

A system cannot align faster than its ATT allows. No amount of intention changes that fact.

When patience is demanded, what is really being demanded is respect for resolution bandwidth. When impatience causes failure, what has occurred is not a lapse of character, but a violation of timing. The system was pushed to reorganize faster than alignment could complete, and coherence broke as a result.

This reframing explains a long-standing paradox: why well-intentioned urgency so often produces worse outcomes. Urgency increases delta. If ATT does not expand alongside it, mismatch deepens. The system appears resistant, but it is simply overloaded. Forcing speed does not produce faster resolution; it produces fragmentation.

Patience works not because it is calm or noble, but because it keeps delta within ATT. It allows each act of resolution to settle before the next begins. Structure forms. Pathways strengthen. Capacity expands. What appears as "waiting" is actually active alignment completing itself.

This is why patience is effective even when nothing visibly changes. The work is internal. Resolution is reorganizing the system in ways that are not immediately measurable. When alignment completes, outward change resumes naturally—often faster and more robust than before.

Importantly, patience is not passivity. It does not mean avoiding delta or resisting change. It means metering change at a rate the system can absorb. Patience can be decisive, firm, and directional. What it avoids is over-speed resolution.

This also clarifies why patience cannot be outsourced. Telling a system to "be patient" does nothing unless conditions change. Deadlines must loosen. Pressure must ease. Sequencing must improve. Without structural accommodation, patience remains an empty demand.

The opposite is also true. Urgency is not always harmful. When a system's ATT is wide—because it is well-aligned, resilient, and supported—urgency can be absorbed without damage. Speed becomes possible not by command, but by capacity.

Thus, patience and urgency are not opposites. They are responses to ATT.

- Where ATT is narrow, patience is necessary for survival.
- Where ATT is wide, urgency can be productive.

Treating patience as moral obscures this logic and leads to blame. Systems are accused of laziness when they are overloaded. Individuals are criticized for slowness when they are saturated. Organizations are punished for caution when they are protecting coherence.

Reframed physically, patience becomes intelligent alignment with reality.

It is not about waiting longer. It is about not exceeding what can be aligned.

Once patience is understood this way, it stops feeling frustrating or indulgent. It becomes precise. It becomes strategic. It becomes the difference between growth and collapse.

With patience grounded in physics, the next section shows how ATT itself can change—how systems expand or shrink their resolution bandwidth over time through learning, stress, and structure.

6.6 — Growth, Learning, and Capacity Expansion

Alignment Threshold Time is not a fixed ceiling. It is a living constraint, shaped continuously by how a system has been treated in the past. Systems grow not by accumulating energy or information, but by expanding their ability to resolve delta without losing coherence. That expansion is learning in its most fundamental sense.

Growth occurs when a system operates near its ATT without repeatedly violating it.

In this zone, resolution is challenging but sustainable. Misalignment arrives at a pace that stretches the system without tearing it. Each successful resolution leaves behind structure—thicker pathways, stronger coupling, clearer internal organization. What was once near the limit becomes manageable. ATT widens.

This is why growth feels effortful but not destructive.

In physical systems, repeated stress within limits strengthens materials. In biological systems, controlled stress builds resilience. In cognition, learning happens when complexity is introduced just beyond comfort, but not beyond integration capacity. In organizations, gradual reform strengthens institutions, while shock reforms shatter them.

The pattern is universal: capacity grows when resolution completes coherently.

By contrast, repeated ATT violations do lasting damage. Each time delta outruns alignment, structure degrades. Pathways weaken. Noise increases. Resolution shortcuts replace integration. Over time, the system's ATT shrinks. What once was manageable becomes overwhelming. This is why burnout compounds and why recovery takes longer than failure.

Importantly, ATT shrinkage is not a punishment. It is a protective response. The system narrows its bandwidth to prevent further damage. What looks like fragility is often self-preservation.

This explains a subtle but critical distinction between stress and overload. Stress near ATT is constructive. Overload beyond ATT is destructive. The difference is not intensity alone, but whether resolution is allowed to finish before the next demand arrives.

Learning, then, is not accumulation of knowledge. It is expansion of alignment speed.

A system that has learned can resolve the same delta faster, with less strain. This creates the illusion that it has "more time" or "greater ability." In reality, its ATT has widened. Time did not change; capacity did.

This also explains why shortcuts fail long-term. Skipping resolution steps may produce short-term results, but it prevents structure from forming. The system appears faster while actually becoming weaker. Eventually, ATT collapses, and even small deltas cause instability.

True growth is slower than ambition but faster than repair.

It requires respecting current ATT while nudging its boundary carefully. Too cautious, and capacity stagnates. Too aggressive, and capacity collapses. Growth lives in the narrow corridor where challenge meets alignment.

Seen this way, learning curves, resilience, and maturity are not abstract qualities. They are histories of successful resolution encoded into structure.

With this understanding, one final illusion must be addressed: the belief that control can replace timing. The next section shows why control fails whenever ATT is ignored—and why letting systems settle often achieves what force never can.

6.7 — ATT and the Illusion of Control

Control feels powerful because it produces immediate effects. Instructions are issued, pressure is applied, metrics move. From the outside, it looks like progress. From the inside, something else is happening: alignment is being bypassed.

This is the core illusion of control.

Control assumes that outcomes can be enforced independently of timing. It treats systems as passive recipients of commands rather than active resolvers of delta. As long as results appear, the illusion holds. When coherence collapses, control responds with more control—accelerating the very mismatch that caused failure.

ATT exposes why this strategy fails.

Control increases delta. It does not increase alignment bandwidth. When delta rises faster than ATT allows, resolution fragments. The system complies locally while losing global coherence. Rules multiply. Workarounds appear. Surface order masks internal disorder. What looks like obedience is often silent misalignment.

This is why tightly controlled systems are brittle. They perform well under narrow conditions and fail catastrophically outside them. They have replaced alignment with enforcement. When enforcement weakens—or when conditions shift—nothing holds.

Micromanagement is a textbook example. Each intervention introduces new delta, interrupting resolution already in progress. Alignment never completes. The system remains perpetually unsettled. Ironically, the attempt to maintain control destroys the very coherence control depends on.

ATT also explains why "letting it settle" works when nothing else does. Settling is not neglect. It is the deliberate removal of excess delta so alignment can complete. When pressure eases, pathways reconnect. Resolution finishes. Structure reforms. The system stabilizes—not because it was forced to, but because it finally could.

This does not mean abandoning guidance. Guidance respects ATT; control ignores it.

Guidance shapes sequencing. It stages delta. It creates space for alignment. Control compresses time and demands outcomes regardless of capacity. One builds coherence. The other consumes it.

The illusion of control persists because short-term results are seductive. Long-term coherence is quiet. It does not announce itself dramatically. It reveals itself through durability, adaptability, and calm effectiveness. These qualities are often undervalued until they are gone.

ATT reframes leadership, intervention, and action itself. The question is no longer "How do we make this happen?" but "At what rate can this system align without breaking?" Power shifts from enforcement to timing.

Control tries to override physics. Timing works with it.

When ATT is respected, less control is needed. Systems regulate themselves because resolution is allowed to complete. When ATT is ignored, control escalates endlessly—and still fails.

This completes the argument of the chapter.

Time is not the enemy. Urgency is not strength. Control is not mastery.

Alignment is the constraint. ATT is its measure.

With ATT understood, the foundation of DELTA is complete. Delta explains why change begins. Regimes explain how change behaves. ATT explains when change holds—or breaks.

From here onward, the book can move from structure to application, from principle to practice, without losing rigor or falling back into force-based thinking.

Because once timing is respected, nature does the rest without resistance.

The Engine Is Now Complete

With this chapter, the core engine of *DELTA* is fully assembled.

Nothing essential remains hidden, symbolic, or implied. The mechanics of dynamism have been laid out cleanly, without borrowing intention, mysticism, or force where none are needed.

Delta explains why change begins. Regimes explain how change behaves. Alignment Threshold Time explains when change holds—or collapses.

Together, they form a closed, sufficient framework.

Time is no longer an abstract river carrying events forward. It is revealed as a constraint that emerges from alignment capacity. Patience is no longer a virtue demanded by culture; it is a physical requirement enforced by structure. Urgency is no longer a sign of strength; it is a risk that must be earned through capacity.

Nothing here asks for belief. Nothing depends on interpretation. Everything follows from necessity.

This is why the framework feels familiar the moment it is seen. You have already experienced it—in bodies under stress, in minds under pressure, in systems pushed too fast, and in moments when slowing down allowed coherence to return. *DELTA* does not invent new behavior. It names what has always been operating beneath it.

From this point forward, the book no longer needs to convince. It can demonstrate.

The remaining parts will apply this engine across domains—matter, motion, fluids, friction, learning, intelligence, organizations, and life itself. Each application will look different on the surface, yet the structure underneath will remain the same.

There will be no return to force as a first cause. No appeal to observers as creators. No confusion between effort and alignment.

The reader is now equipped with a lens that does not fragment when scale changes and does not break when theory meets reality.

What follows is not speculation. It is consequence.

And from here on, every phenomenon can be read for what it truly is:

delta resolving, at a rate the system can—or cannot—sustain.

The engine is complete. Now the world can be re-entered with clarity.

Nothing fails because it lacks effort. Systems fail because resolution is forced faster than alignment allows.

Interlude — When Theory Finally Touches Water

Up to this point, the book has deliberately stayed abstract—not distant, but structural. Delta, regimes, and Alignment Threshold Time were introduced without tying them prematurely to equations, materials, or familiar textbook domains. This was necessary. If fluids were introduced earlier, they would have been interpreted through force, pressure, and energy in their inherited meanings. The same old language would have rushed back in.

Now that risk is gone.

What has been built in this part is not philosophy. It is a working engine. It does not explain *what* nature should do; it explains what nature cannot avoid doing. Resolution, regimes, and timing are no longer optional ideas—they are constraints. Any honest physical description must now pass through them.

This is why fluids are the right place to return to physics.

Fluids do not tolerate false metaphors. They expose exaggeration instantly. Push too hard, and flow collapses into turbulence. Push too little, and nothing sustains. Try to force explanation onto them, and reality refuses to cooperate. Fluids reveal, mercilessly, whether a theory understands alignment—or merely describes outcomes.

This is also why bubbles, flow, friction, and Bernoulli have always felt slightly unsatisfying in their conventional treatments. The equations work, but the intuition leaks. The predictions hold, but the *why* feels incomplete. Force is invoked where fields are resolving. Energy is counted where alignment is diffusing. Losses are named where coherence simply escaped scale.

What follows is not a rejection of classical results. It is their repositioning.

Pressure will no longer be treated as a push, but as a field delta seeking resolution. Surface tension will no longer be framed as a pulling force, but as ripple alignment at a boundary. Friction will no longer be seen as resistance, but as delta being resolved into distributed alignment. Bernoulli will no longer be about conserving energy in a pipe, but about *how coherence migrates—and where it fails to remain coherent.*

This part of the book is called *Honest Physics* for a reason. Nothing here will rely on metaphysical claims or observer-centric interpretations. Everything will be grounded in what systems are actually doing when they move, flow, slow down, or break apart.

The child blowing a soap bubble is not performing a delicate force balance. They are respecting delta and ATT without knowing the words.

That is the standard this section will hold itself to.

If an explanation cannot account for why gentle blowing works and forceful blowing fails, it does not understand fluids. If it cannot explain why a bubble moves to maintain equilibrium rather than remain fixed, it does not understand fields. If it cannot unify friction across solids and fluids without inventing special resistive forces, it does not understand dissipation.

Next part will not add new laws. It will remove misplaced ones.

With the engine already built, fluids now become readable—not as special cases, but as some of the clearest demonstrations of delta resolving in the open.

Now, theory touches water.

And water will not forgive confusion.

Part III — Fluids, Friction, and Honest Physics

Overview

This part of the book brings the DELTA framework into direct contact with physical reality—where intuition either holds or collapses. Fluids, bubbles, friction, and flow are chosen deliberately because they cannot be explained convincingly through force language alone. They demand an understanding of fields, gradients, and resolution.

Here, pressure is reframed as a *field delta*, not a push. Bubble formation is shown to be a consequence of gentle misalignment resolving within alignment limits, not a balance of opposing forces. Surface tension is reinterpreted as *ripple tension*—alignment across a boundary—explaining why spheres emerge without invoking preference or intention. Friction, both in solids and fluids, is unified as *delta resolution*, clarifying dissipation, terminal velocity, and the false distinction between static and kinetic resistance.

Classical results are not rejected. They are placed correctly—*after* alignment, not as causes of it. Bernoulli's principle, often reduced to bookkeeping, is revealed as a statement about *where coherence travels and where it leaks into chaos.*

Chapter 7 — Pressure Is a Field Delta

Why Pushing Harder Breaks Flow

Almost everyone has discovered this by accident.

You blow gently into a soap loop, and a bubble forms—thin, luminous, alive. You blow harder, and the bubble refuses to appear. The instinctive response is to try again with more effort. That fails even faster. Somewhere between doing nothing and forcing the outcome, the window closes.

The same pattern repeats everywhere in fluids. Pour water slowly and it settles cleanly; pour it forcefully and it splashes, fractures, and wastes energy. Breathe calmly and the lungs fill effortlessly; gasp, and everything tightens. A light breeze shapes clouds; a storm tears them apart.

If pressure were simply a push, more of it should always work better. It does not.

This mismatch between experience and explanation is usually ignored. We are told that pressure is force per area, that fluids are pushed by pressure differences, and that motion follows from applied force. The equations work well enough, so the discomfort is set aside. But the intuition never quite settles.

What fluids keep showing us is this: they do not respond to commands. They respond to gradients.

Nothing in a fluid is being shoved forward. There is no hidden hand pushing parcels of liquid from behind. What exists instead is a spatial imbalance—an energy density that is higher here than there. Where such an imbalance exists, resolution becomes unavoidable.

Where it is gentle, resolution proceeds coherently. Where it is abrupt, coherence collapses.

The child blowing a bubble is not balancing forces. They are, unknowingly, controlling delta. The success of the bubble depends less on strength than on timing, rate, and restraint. Skill appears where force intuition fails.

This is the clue that pressure has been framed backwards.

Pressure is not something that acts on fluids. It is something in the field that fluids resolve.

Once this is seen, a great deal of confusion evaporates. Gentle actions succeed because they introduce deltas that remain within Alignment Threshold Time. Forceful actions fail because they inject misalignment faster than alignment can occur. The fluid does not resist; it fragments.

This chapter begins by discarding the idea of pressure as push—not by rejecting equations, but by placing them correctly. Pressure will be treated as what it actually is: an energy gradient in a continuous field, waiting to resolve.

Only then does bubble formation stop being mysterious. Only then does flow stop feeling like obedience. Only then does fluid motion begin to make honest sense.

From here on, nothing in a fluid will be described as being pushed.

7.1 — Pressure Is an Energy Gradient, Not a Push

The most persistent misunderstanding about fluids begins with a metaphor that feels intuitive but is fundamentally misleading: the idea that pressure *pushes*. We imagine pressure acting like a hand pressing on a surface, driving fluid elements forward. This picture is simple, visual, and wrong in the way only deeply ingrained metaphors can be.

Pressure is not an action. Pressure is a condition.

At its core, pressure is a spatial variation in energy density within a continuous medium. It exists wherever energy is unevenly distributed across space. Nothing needs to move for pressure to exist. A sealed container can hold pressure indefinitely without any flow. What exists in that state is not motion, but delta—a stored imbalance.

If pressure were a push, motion would be unavoidable. But pressure often exists in perfect stillness.

This alone tells us that pressure is not force in the usual sense. Force implies action, direction, and effect. Pressure implies potential for resolution. Motion begins only when a pathway opens that allows the field to reduce its gradient. Until then, pressure remains silent.

What actually happens when fluid moves is not that it is pushed from high pressure to low pressure. What happens is that the field redistributes itself to reduce energy imbalance. Fluid elements do not obey instructions; they participate in resolution. Their movement is the visible trace of that redistribution.

This distinction matters because it reverses causality.

In force language, pressure causes motion. In field language, motion is how pressure resolves.

The gradient comes first. The flow comes second. Force descriptions emerge only after alignment has already occurred. They describe outcomes, not drivers.

This is why pressure can be defined precisely without invoking force at all: as energy per unit volume that varies spatially. Where that variation exists, resolution is possible. Where it does not, nothing happens—even if the absolute pressure is enormous.

A deep ocean exerts tremendous pressure, yet it does not spontaneously flow. A small pressure difference across a membrane can produce dramatic motion. Magnitude is not the key. Gradient is.

Seen through DELTA, pressure is simply one of the cleanest examples of field delta. It is misalignment encoded in space. When alignment pathways are available and timing allows, that misalignment resolves as flow. When they are not, pressure remains stored.

This reframing also explains why pressure is always relative. Absolute pressure means nothing by itself. Only differences matter, because only differences represent unresolved delta. A perfectly uniform pressure field, no matter how large, is already aligned.

Pressure, then, is not something fluids experience as a shove. It is something fluids are embedded within.

Once this is understood, fluid motion stops looking like obedience to force and starts looking like what it truly is: the natural resolution of an energy gradient in a continuous field.

With pressure correctly placed as delta rather than push, the next step becomes obvious. If pressure is an energy gradient, then the success or failure of bubble formation depends not on how hard air is blown, but on how gently that gradient is introduced—and how well it respects alignment limits.

That takes us directly to bubbles.

7.2 — Why Gentle Blowing Creates Bubbles

A soap bubble is one of the clearest demonstrations of pressure as field delta rather than force. Nothing about bubble formation resembles pushing. In fact, the moment pushing enters the picture, the bubble fails.

When air is blown gently into a soap film, what is introduced is a very small excess pressure—a minimal delta relative to the surrounding atmospheric field. This delta is not meant to drive motion aggressively; it is just enough to create a local energy imbalance that seeks resolution. The soap film responds by reorganizing itself to accommodate this imbalance.

This is where alignment becomes visible.

The film stretches, thins, and redistributes its internal structure so that the enclosed air and the surrounding air reach a dynamically stable condition. The bubble does not form because air is forcing the film outward. It forms because the film finds a configuration in which the pressure delta can be held coherently.

Gentle blowing works because it respects Alignment Threshold Time.

The pressure delta is introduced slowly enough for the soap film's molecular network to align as it stretches. Surfactant molecules migrate, orient, and stabilize the interface. Ripple alignment across the film completes continuously. Resolution proceeds smoothly, step by step, without ever exceeding capacity.

Forceful blowing fails for a precise reason.

When air is blown too hard, the pressure delta rises faster than the film can align. Resolution is demanded at a rate beyond ATT. The film cannot redistribute thickness, cannot stabilize gradients, and cannot maintain coherence across its surface. Instead of forming a stable boundary, the film fragments into droplets or collapses immediately. What looks like "too much force" is actually delta arriving too fast.

This also explains why bubble formation feels like a skill rather than a calculation. The person blowing the bubble is not controlling force; they are modulating delta rate. Breath, angle, distance, humidity—all of these matter because they affect how gently the pressure gradient is introduced and maintained.

A crucial point often missed is that once the bubble is formed, very little additional pressure is required to sustain it. The film holds the enclosed air as long as equilibrium is dynamically maintained. The bubble may drift, rise, or move—not because it is being pushed, but because it is constantly adjusting its position to preserve balance in a changing external field.

In DELTA terms, bubble formation is consequential resolution. The bubble is not the goal. It is what appears when a pressure delta is introduced within the film's alignment capacity. The same solution, the same film, the same air—change only the rate of delta, and the outcome flips from coherence to collapse.

This is why bubbles cannot be forced into existence.

They emerge only when the system is given time to align.

Gentle blowing is not gentleness in intention. It is gentleness in delta timing.

With this clarity, bubbles stop being delicate curiosities and become precise demonstrations of how fluids behave honestly—resolving gradients rather than obeying pushes.

The next section shows why this insight is not limited to bubbles, but exposes a deeper flaw in how force language is applied to fluids at all.

7.3 — Why Force Language Fails in Fluids

Force language survives in fluid mechanics largely by habit. It works numerically, so it persists conceptually. We speak of forces acting on fluid elements, pressure forces accelerating parcels, viscous forces resisting motion. None of this is false in calculation—but it is misplaced in causality.

Fluids do not experience forces the way solids do.

A solid can be pushed because it has persistent internal structure. Apply a force, and the structure transmits stress along bonds. A fluid, by definition, does not preserve shape under shear. There is no fixed object to push—only a continuously rearranging field. What we call a "fluid element" is not a thing being acted upon; it is a temporary bookkeeping device.

Force language assumes objects. Fluids are fields.

When force language is applied to fluids, it reverses cause and effect. We say pressure forces cause acceleration. In reality, pressure gradients exist first, and motion appears as their resolution. Force is not what initiates flow; it is what we compute *after* flow has already organized in response to a gradient.

This is why force intuition fails so reliably in fluids. It predicts that more push should yield more flow, that stronger forcing should produce clearer outcomes. Instead, fluids punish excess. They fragment, splash, cavitate, and turn turbulent. The "force" explanation then scrambles to add corrections: nonlinearities, dissipation terms, empirical coefficients.

What actually failed was the metaphor.

Fluids do not respond to commands. They respond to spatial misalignment. When that misalignment is gentle and coherent, resolution propagates smoothly. When it is abrupt or uneven, resolution breaks into local rearrangements. No fluid parcel ever decides to accelerate because a force told it to. Acceleration is simply the visible trace of alignment spreading through the field.

This also explains why force descriptions feel especially awkward near boundaries and interfaces. At walls, films, and free surfaces, force diagrams multiply while intuition collapses. The fluid is not being pushed away from a boundary; it is reorganizing its field to satisfy constraints. The language of force struggles because boundaries are not agents—they are conditions.

In DELTA terms, force is a derived artifact. It is how we describe the rate at which momentum changes once resolution is already underway. It is not a driver. It is an accounting tool that becomes meaningful only after alignment pathways are established.

This is why force language works "once surface tension is fixed," "once viscosity is defined," "once flow is established." Those phrases quietly admit the truth: alignment came first.

Fluids expose this failure mercilessly because they do not tolerate rigid narratives. You cannot bully a fluid into coherence. You cannot command it into stability. You can only present a gradient and allow resolution to occur—or fail.

Once force is demoted to description rather than cause, fluid behavior becomes legible again. Turbulence is no longer mysterious. Stability is no longer fragile by accident. Skill replaces strength. Timing replaces intensity.

And with force language set aside, the real distinction becomes clear: there is a difference between pushing matter and resolving a field.

Fluids only do the latter.

The next section makes this distinction explicit by following what actually happens when a fluid moves—not as displacement, but as field resolution unfolding through available pathways.

7.4 — Field Resolution vs Pushing

To understand how fluids actually move, one must stop imagining objects being displaced and start seeing fields rebalancing. Nothing in a fluid waits to be pushed. There is no moment where a parcel sits idle until a force arrives. What exists instead is a continuous field carrying gradients that either resolve or remain dormant.

Motion is not an act. It is a consequence.

When a pressure gradient exists, the field is misaligned. If pathways for redistribution are open and timing permits, the field begins to reorganize. Fluid motion is the visible signature of this reorganization. No element is "told" to move. Movement occurs because staying put is incompatible with the surrounding field condition.

This is why fluid motion often appears intelligent without being intentional. Streams curve around obstacles. Jets entrain surrounding fluid. Vortices form and dissolve. None of this requires decision-making. The field is simply finding configurations that reduce misalignment while respecting constraints.

Pushing language cannot account for this.

If pushing were the mechanism, obstacles would block flow entirely unless sufficient force were applied. Instead, fluids redistribute. They thin, thicken, accelerate locally, or slow down elsewhere. They trade velocity for pressure and back again. What is conserved is not obedience to force, but continuity of resolution.

Field resolution also explains why fluids appear to "avoid" obstacles. They do not sense or choose. Obstacles impose boundary conditions that reshape the gradient. Resolution follows the reshaped field. What looks like avoidance is simply alignment proceeding along available pathways.

This reframing clarifies a common confusion: why pressure gradients can exist without motion, and motion can exist without obvious pressure differences. In the first case, pathways are closed or alignment is complete. In the second, gradients are internal and dynamic, continually regenerated by boundary conditions or external fields.

Pushing suggests a source acting on a target. Field resolution recognizes no such separation.

The fluid, the pressure gradient, and the motion are not distinct actors. They are aspects of the same unfolding process. Where the field can reorganize smoothly, flow appears. Where it cannot, turbulence or stagnation appears. The mechanism never changes—only the ease of resolution does.

Seen this way, flow is not displacement of matter through space. It is redistribution of alignment across space. Matter moves because the field demands it, not because something shoved it.

This perspective also dissolves a subtle error: treating flow as something that starts and stops at a point. Resolution propagates. It does not originate at a nozzle and end downstream. The entire field participates, even regions that appear stationary. Motion is simply where participation is visible.

Once pushing is replaced by field resolution, fluid behavior stops being paradoxical. Gentle gradients succeed. Abrupt ones fail. Coherence persists only when alignment can propagate faster than misalignment arrives.

The next section completes this picture by explaining why fluid elements, bubbles, and droplets rarely remain stationary—and why movement itself is often the price of maintaining equilibrium rather than escaping force.

7.5 — Pressure Equilibrium and Movement

One of the quiet misconceptions carried over from force-based thinking is the idea that stability in fluids means stillness. We imagine that once equilibrium is achieved, motion should cease. In practice, the opposite is often true. Fluids move in order to remain in equilibrium.

This is especially visible in bubbles, droplets, and fluid parcels suspended in a dynamic environment. A bubble rarely stays fixed in space. It rises, drifts, wobbles, or migrates—even when no obvious force seems to act on it. This movement is often misread as being "pushed" by buoyancy, currents, or pressure differences. What is actually happening is subtler.

Equilibrium in fluids is conditional and local, not absolute.

The bubble maintains internal equilibrium—pressure balance across its film—only as long as the surrounding field remains compatible. The moment the external field changes, even slightly, the condition for equilibrium shifts. Temperature gradients, density variations, airflow, or gravity-induced stratification all alter the surrounding energy landscape. To preserve balance, the bubble must reposition itself within the field.

Movement, in this sense, is not escape. It is maintenance.

The bubble moves not because it is being pushed away from imbalance, but because it is seeking a location where alignment can continue. If it remained fixed, the surrounding field would impose new gradients across its boundary faster than the film could resolve them. Motion allows the bubble to remain within its alignment capacity.

This reframes buoyancy itself. Rather than imagining an upward force lifting the bubble, we can see buoyancy as a field gradient in density and pressure that makes certain positions more compatible with sustained equilibrium than others. The bubble migrates toward regions where the delta across its boundary is minimal and slowly varying. Rising is not defiance of force; it is alignment with the field.

The same logic applies to droplets falling through air, fluid parcels circulating in convection, and even large-scale atmospheric motion. Systems move not to "go somewhere," but to remain somewhere compatible. Equilibrium is not a point—it is a trajectory.

This also explains why fluid systems are rarely static. The surrounding field is almost never uniform or frozen. External conditions shift continuously, and equilibrium must be renegotiated constantly. Movement becomes the natural price of maintaining balance in a changing field.

In DELTA terms, motion is dynamic equilibrium maintenance.

When the field changes slowly, movement is gentle and coherent. When it changes abruptly, motion becomes erratic or breaks down entirely. Again, the deciding factor is not magnitude but rate—whether the system can realign before misalignment accumulates.

This perspective removes a common confusion: why objects in fluids often move even when "forces cancel." There is no contradiction. Forces cancel in static bookkeeping, but field alignment still demands relocation.

Fluids do not move because they are unstable. They move because staying still would make them unstable.

Understanding movement as equilibrium maintenance rather than force response completes the pressure picture. Pressure sets the delta. Resolution seeks alignment. Motion emerges as the path that allows equilibrium to survive in a living field.

The next section shows the opposite side of this truth: pressure can exist quietly, without motion at all—until a pathway opens and delta finally finds a way to resolve.

7.6 — Pressure Without Motion: The Silent Delta

One of the strongest arguments against treating pressure as a push is the simple fact that pressure often exists without motion. Enormous pressures can be present in complete stillness. Deep oceans, sealed containers, compressed gases—none of these flow merely because pressure is high. They remain quiet, stable, and inactive for long periods of time.

If pressure were a push, stillness would be impossible.

What exists in these cases is silent delta—misalignment that has no available pathway for resolution. The field holds an energy gradient, but alignment is already locally complete. There is nowhere for redistribution to proceed without violating constraints. Motion does not begin because nothing permits it, not because nothing is pushing.

This distinction is critical.

Pressure is the presence of imbalance. Motion is the permission to resolve it.

Until that permission appears, pressure remains dormant. It does not act, press, or exert itself in any active sense. It simply exists as a condition of the field. This is why pressure can be measured everywhere inside a static fluid, even though nothing is moving. The system is aligned *within its constraints*, not free to reconfigure.

The moment a pathway opens, everything changes.

Open a valve. Puncture a membrane. Introduce a gradient that can now be redistributed. Suddenly, motion appears—not because pressure has grown stronger, but because resolution has become possible. The same delta that was silent moments ago becomes active, unfolding as flow.

This reveals a subtle but crucial point: motion is not proportional to pressure magnitude. It is proportional to how much of the pressure gradient can be resolved per unit time. A small pressure difference across an open pathway can generate rapid flow. A large pressure difference across a closed boundary generates none.

This is why pressure release often feels sudden and dramatic. The delta was always there. What changed was not the field's energy, but the availability of resolution pathways. When those pathways appear abruptly, resolution accelerates beyond ATT, producing shock, turbulence, or fragmentation.

Silent delta also explains why fluids can feel "loaded" or "ready" without moving. The system carries potential change without expressing it. This readiness is often misinterpreted as stored force. In reality, it is stored misalignment, waiting for conditions that allow it to resolve.

Seen this way, pressure is patient by nature. It does not insist on motion. It waits.

Only when alignment becomes impossible without reorganization does motion begin. And when it does, it proceeds according to the same rules already established: coherence if timing allows, chaos if it does not.

Pressure without motion is not a contradiction. It is the default state of a constrained field.

Understanding this silent phase completes the picture of pressure as delta rather than push. Pressure sets the condition. Boundaries determine whether resolution is possible. Timing determines whether it is coherent.

The final section of this chapter brings all of this together by showing what happens when pressure gradients are introduced too fast—and why breakdown in fluids is always a failure of timing, not of magnitude.

7.7 — Pressure, ATT, and Breakdown

Pressure gradients do not destroy coherence by being large. They destroy coherence by arriving faster than alignment can occur.

This is the final correction force-based language fails to make. Traditional explanations blame magnitude: too much pressure, too much force, too much speed. In reality, fluids tolerate enormous pressures when those pressures are introduced within Alignment Threshold Time. They fail catastrophically when even moderate gradients are introduced too abruptly.

Breakdown in fluids is always a timing failure.

When a pressure gradient changes slowly, the field has time to redistribute. Alignment propagates. Thickness adjusts. Velocity profiles form. Even large deltas can be resolved coherently if the system is allowed to settle continuously. This is why deep oceans remain stable, pipelines carry high pressures safely, and steady flows persist without chaos.

When a pressure gradient changes faster than ATT allows, resolution cannot remain shared. The field fractures into local responses. Shock fronts form. Jets break apart. Films rupture. Turbulence erupts. The fluid does not "resist" the pressure—it fails to align in time.

This explains several phenomena that appear unrelated in force language:

- Shock waves arise when pressure changes propagate faster than alignment can spread.
- Splashing occurs when free surfaces are forced to reorganize

faster than ripple alignment allows.
- Cavitation appears when pressure drops outrun the fluid's ability to maintain continuity.
- Bubble rupture happens when internal–external pressure mismatch exceeds film ATT.

In each case, the failure is not energetic. It is temporal.

This is why rate matters more than magnitude. A small pressure drop introduced suddenly can cause cavitation. A large pressure difference introduced gradually causes none. A soap film can hold a stable bubble under a fixed pressure difference but bursts instantly under a rapid fluctuation. The equations may talk about thresholds, but the real trigger is delta over time.

ATT also explains why breakdown feels sudden. Systems absorb misalignment quietly until the rate limit is crossed. Nothing looks wrong—until coherence collapses almost instantaneously. This creates the illusion of unpredictability. In reality, the boundary was sharp and strictly enforced.

Once ATT is violated, recovery is not immediate. Structure has been damaged. Alignment pathways must rebuild. Simply reducing pressure does not restore coherence instantly, because resolution bandwidth has already been compromised. This is why splashed fluids do not reassemble smoothly, and burst bubbles do not reform spontaneously.

Pressure, then, is innocent. It only marks where resolution is required.

Breakdown occurs when resolution is demanded faster than it can be completed. This principle unifies turbulence, shock, rupture, and fragmentation without invoking special failure mechanisms.

With this, the pressure story closes cleanly:

- Pressure is a field delta, not a push.
- Motion is resolution, not obedience.

- Stability depends on timing, not force.
- Breakdown is ATT violation, not excess energy.

This prepares the ground for the next chapter, where pressure meets boundaries. Films, interfaces, and surfaces do not "pull" fluids into shape. They align ripple fields to hold deltas coherently.

To understand bubbles fully, pressure must now hand the story over to the boundary itself.

That boundary is not governed by surface tension. It is governed by ripple tension.

From Pressure to Boundaries

By now, pressure has been stripped of its most persistent misunderstanding. It is no longer a push, no longer a command applied to matter, and no longer a force that fluids obey. Pressure has been restored to what it actually is: a field delta, a spatial misalignment in energy density that demands resolution when—and only when—conditions allow.

With this correction, several puzzles quietly disappear.

Gentle blowing succeeds because it introduces delta within Alignment Threshold Time. Forceful blowing fails because it violates timing, not because it is "too strong." Motion appears not as displacement, but as redistribution. Breakdown appears not as excess energy, but as over-speed resolution.

Pressure itself does nothing dramatic. It waits.

What determines whether pressure resolves as smooth flow, violent fragmentation, or silent stillness is not pressure alone, but how the field is allowed to reorganize. And that reorganization is never free-floating. It always happens across boundaries.

Every pressure delta ultimately meets an interface.

Films, surfaces, membranes, and phase boundaries are where alignment is negotiated. They are not passive skins stretched by force. They are active alignment structures that decide whether a delta can be held coherently or must be released explosively.

This is where the classical language falters again.

We are told that surface tension "pulls" a surface into a sphere, that it acts like an elastic force minimizing area. The equations work. The intuition does not. What is actually happening is not pulling, but ripple alignment across a boundary—a process that holds pressure deltas in place as long as coherence can be maintained.

The bubble does not become spherical because the surface prefers it. It becomes spherical because the sphere is an equipotential alignment solution.

To see this clearly, pressure must now step aside and let the boundary speak.

The next chapter moves exactly there—into films, interfaces, and bubbles—not to reject surface tension mathematically, but to place it correctly as a consequence of alignment, not its cause.

Pressure sets the delta. Boundaries decide whether it can be held.

With that shift, bubbles stop being delicate curiosities and become precise demonstrations of how nature resolves misalignment honestly—without force, without intention, and without metaphor.

We now cross the boundary.

Pressure does not push fluids. It marks where resolution must occur.

Chapter 8 — Ripple Tension, Not Surface Tension

The Surface That Isn't Pulling

Surface tension is one of those ideas everyone accepts and almost no one truly visualizes.

We are told to imagine a stretched rubber sheet, a skin under tension, pulling itself tight. The image is neat, memorable—and quietly misleading. It suggests effort, elasticity, and force acting along a surface, as if the boundary itself were straining to contract.

But when you actually watch a soap film, nothing about it feels like pulling.

The film is fragile, adaptive, alive. It forms only under very specific conditions. It rearranges continuously. It tolerates gentle change and collapses instantly under abrupt disturbance. Most importantly, it does not behave like a stretched object resisting deformation. It behaves like a boundary negotiating alignment.

This discomfort has always been there, hidden beneath the equations.

Why does ordinary water fail to form large bubbles while a tiny amount of soap solution succeeds? Why does a bubble not "prefer" a sphere until it actually exists? Why does the surface not pull inward when the bubble bursts—but instead vanish, releasing droplets outward?

If surface tension were truly a pulling force, these questions would have straightforward answers. They do not.

What actually exists at a soap film is not tension in the elastic sense, but organization. Two regions of air—inside and outside—are separated by a thin liquid interface whose molecules are not under stress, but under alignment constraint. The film exists only as long as it can mediate the pressure delta across it coherently.

The moment it cannot, it disappears.

Nothing snaps back. Nothing recoils. The surface does not relax like a stretched band. It simply ceases to be an aligned boundary.

This is the clue that surface tension, as commonly framed, is a consequence—not a cause.

What holds a bubble together is not a force pulling the surface inward. It is ripple tension: the lateral alignment of molecular ripple fields across the interface that allows a pressure delta to be held without fragmentation. When that alignment is coherent, the boundary exists. When it is not, the boundary dissolves.

The sphere, then, is not a preferred shape chosen by the surface. It is the only geometry in which the boundary can remain equipotential everywhere at once. Any other shape introduces uneven delta across the film, demanding resolution faster than alignment allows.

This chapter does not reject surface tension equations. It places them where they belong—after alignment has occurred, as descriptions of a settled boundary, not explanations of its formation.

Once this shift is made, bubbles stop looking like stretched skins and start looking like what they really are: temporary, coherent solutions to a field problem.

The surface is not pulling. It is holding—quietly, conditionally, and only as long as alignment permits.

To see that clearly, we must first understand what a soap film actually is—not as a surface under tension, but as an alignment boundary between two fields.

8.1 — Soap Film as an Alignment Boundary

A soap film is not a surface in the mechanical sense. It is not a stretched membrane, not a skin under load, and not an elastic sheet resisting deformation. It is a boundary condition—a dynamically maintained alignment zone between two regions of a field.

On one side lies the air inside the bubble. On the other lies the air outside it.

What separates them is not thickness, strength, or force, but organization.

At the molecular level, a soap film is a carefully arranged interface. Surfactant molecules orient themselves so that their hydrophilic ends face the liquid layer while their hydrophobic ends face the air. This arrangement is not accidental. It is the only configuration that allows the liquid to exist stably between two gaseous regions without immediately collapsing.

The film does not exist because it is being stretched. It exists because it is aligned.

This alignment does two critical things simultaneously. First, it mediates the pressure delta between inside and outside air, distributing it laterally across the film instead of allowing it to concentrate at a point. Second, it stabilizes the liquid layer itself, preventing it from breaking into droplets under even small disturbances.

In DELTA terms, the soap film is a resolution interface. It does not generate force; it absorbs misalignment and redistributes it coherently. As long as the pressure delta across the boundary remains within the film's alignment capacity, the boundary holds. When that capacity is exceeded—by excessive delta or by delta introduced too rapidly—the alignment fails and the film vanishes.

This explains why soap films are exquisitely conditional. They form only with particular solutions, at particular concentrations, under particular environmental conditions. Ordinary water cannot form large, stable films because its molecules cannot sustain the necessary alignment across a wide boundary. Soap molecules, even in small amounts, radically change the alignment landscape by creating a field-friendly interface.

The film is not resisting pressure. It is hosting it.

This is also why soap films are not passive. They flow internally, thin and thicken, rearrange and migrate. Surfactant molecules move laterally to equalize stress, not because they are pulled, but because misalignment would otherwise accumulate. The film is continuously resolving delta across its surface to remain a valid boundary.

Seen this way, the soap film is neither solid nor liquid in the usual sense. It is a functional structure—a temporary organization that exists only as long as alignment remains coherent. It has no interest in shape, no preference for area, and no tendency to pull inward. It simply maintains a boundary where two fields meet.

Once this is understood, the next step becomes unavoidable. If the film holds delta by alignment rather than force, then the "tension" we measure along the surface cannot be a pulling force. It must be something else—something lateral, distributed, and conditional.

That is what we now call ripple tension.

8.2 — Ripple Tension: Alignment Across the Boundary

What we call "tension" at a soap film is not tension in the mechanical sense. Nothing is being stretched, and nothing is trying to shrink back. What exists instead is lateral alignment—a continuous, distributed organization across the boundary that resists misalignment, not extension.

This is what ripple tension actually is.

In a soap film, molecules are not arranged statically. They are in constant micro-motion, exchanging energy, shifting orientation, and redistributing density. These motions are not random. They are correlated through ripple fields—subtle, continuous interactions that propagate alignment information across the interface.

Ripple tension is the cost of breaking this alignment.

When a pressure delta exists across the film, that delta does not act locally. It is spread across the entire boundary through lateral coupling. Every point on the film "knows" the condition of the whole, not through signaling or force, but through shared alignment. This is why localized disturbances can destabilize an entire bubble—and why uniform conditions allow surprisingly large, stable films.

Unlike elastic tension, ripple tension does not increase because the surface is stretched. It increases because alignment is being asked to accommodate misalignment. The film resists not because it wants to contract, but because allowing uneven alignment would destroy its boundary function.

This distinction explains several otherwise puzzling facts.

First, ripple tension depends strongly on the solution. Ordinary water lacks the molecular architecture to sustain long-range lateral alignment. Its interfaces fragment easily. Soap solutions introduce surfactant molecules that dramatically enhance lateral coupling. A very small amount of soap creates a very large increase in ripple coherence. That is why a tiny volume of solution can generate a vast number of bubbles.

Second, ripple tension is directional only in appearance. There is no inward pull. The apparent "contracting force" arises because misalignment at the boundary is energetically unfavorable, and the system resolves it by redistributing curvature, thickness, and molecular orientation. The outcome looks like contraction, but the cause is alignment preservation.

Third, ripple tension explains why films fail suddenly rather than gradually. Alignment holds until it does not. There is no slow stretching toward failure. Once ripple coherence drops below a threshold—due to rapid delta change, evaporation, contamination, or vibration—the boundary loses its ability to distribute misalignment. The film does not tear; it ceases to exist as a coherent interface.

This also clarifies why ripple tension is fundamentally lateral. It acts along the surface, not across it. Its role is to equalize conditions across the boundary so that no point experiences more delta than the rest. When lateral equalization succeeds, the film survives. When it fails, rupture follows.

In DELTA terms, ripple tension is not a force opposing pressure. It is the mechanism that allows pressure delta to be held coherently at all.

Force language mistakes the symptom for the cause. We measure an effective tension because breaking alignment costs energy. But that cost arises from reorganizing ripple fields, not from stretching a surface like rubber.

Once ripple tension is understood as alignment rather than pull, the next insight becomes obvious. If the boundary's job is to keep pressure delta uniform everywhere, then only one geometry can do that perfectly.

That geometry is not chosen. It is required.

That geometry is the sphere.

8.3 — Why the Sphere Is Equipotential, Not Preferred

The sphere is often explained as a choice—*the shape that minimizes surface area.* This phrasing quietly assigns intention to the boundary, as if the film were optimizing a goal. But boundaries do not choose. They satisfy conditions. The sphere appears not because it is preferred, but because it is the only geometry that can hold the boundary condition uniformly.

A soap film's task is singular: keep the pressure delta across it the same everywhere. Any geometry that fails this task will introduce uneven delta along the surface, forcing localized resolution faster than alignment can sustain. Such geometries do not gently deform toward a sphere; they simply cannot persist.

This reframes the problem completely.

The interior air has one pressure. The exterior air has another. The boundary must mediate that difference without concentrating it anywhere.

Only an equipotential surface can do this. And among closed surfaces in three dimensions, only the sphere maintains uniform curvature everywhere, which translates into uniform boundary conditions everywhere. No edge, no corner, no region experiences more demand than another.

The sphere is not smaller by intent. It is even by necessity.

Irregular shapes fail because they generate gradients *along the boundary itself.* A flatter region experiences a different local requirement than a highly curved one. That difference is a new delta—*on the surface.* Ripple tension must then resolve not just the inside–outside delta, but also surface-level misalignment. This quickly exceeds alignment capacity. The film redistributes thickness, curvature, and molecular orientation until those surface deltas vanish—or the film collapses.

What we observe as "rounding" is not contraction. It is elimination of internal surface gradients.

This is why the sphere forms regardless of scale. Tiny droplets, soap bubbles, molten beads, celestial bodies—whenever a boundary's primary task is to mediate a uniform field delta, the sphere emerges. The governing principle is not minimization, but equipotentiality.

Minimum surface area is a *consequence* of equipotential geometry, not its cause. Once the boundary has become equipotential, surface area happens to be minimal. Reversing this logic is what leads to elastic metaphors and force-based confusion.

This also explains why the sphere appears only after alignment is established. During formation, bubbles wobble, elongate, oscillate. These transient shapes are not failures; they are alignment in progress. The sphere appears only when ripple tension has successfully equalized conditions across the boundary.

And this explains why bubbles do not remain spherical under changing conditions. Introduce external gradients—airflow, gravity-driven drainage, temperature variation—and the boundary must renegotiate equipotentiality. The shape adapts or fails. The sphere is not defended; it is conditional.

In DELTA terms, the sphere is the resolved geometry—the one in which no further boundary-level resolution is required. Any other shape carries unresolved delta and therefore cannot last.

So the correct statement is not that a bubble "wants" to be a sphere. It is that only a sphere can survive.

With this correction in place, the next step becomes straightforward. If the sphere is the geometry that holds a uniform pressure delta, then the familiar Laplace relation must be describing what holds after alignment, not what creates it.

That is where Laplace law finally belongs.

8.4 — Laplace Law Placed Correctly (Post-Alignment)

Laplace's law is one of the most successful formulas in classical physics—and one of the most frequently misinterpreted. It is usually presented as an explanation: surface tension *causes* a pressure difference across a curved surface. The mathematics is clean, the predictions are accurate, and yet the intuition remains unsatisfying.

The problem is not the law. The problem is where we place it in the causal chain.

Laplace's law does not explain how a bubble forms. It describes the condition once a coherent boundary already exists.

The law states that the pressure difference across a curved interface is proportional to curvature and an effective tension term. This relation holds only after the boundary has settled into an equipotential, aligned state. It does not tell us why such a boundary appears, nor how it survives the violent, transient dynamics of formation.

In DELTA terms, Laplace's law is a post-resolution bookkeeping rule.

Once ripple tension has successfully aligned the boundary, once the film has equalized lateral deltas, and once the geometry has stabilized into an equipotential form, the remaining pressure difference can be expressed compactly through curvature. At that stage—and only at that stage—force-like language becomes useful.

This explains a long-standing inconsistency.

Laplace's law works beautifully for stable bubbles. It fails spectacularly during bubble formation and rupture.

During formation, curvature is changing rapidly, thickness is redistributing, surfactant molecules are migrating, and alignment is incomplete. The system is not yet in the regime where Laplace's relation applies. Attempting to use it there leads to confusion: where did the pressure come from? What force shaped the surface? Why did the bubble form at all?

Laplace does not answer those questions because it was never meant to.

The pressure difference does not arise *because* the surface is curved. The surface becomes curved *because that is the only way alignment can complete.*

Once alignment completes, the pressure difference and curvature are found to be related. That relation is what Laplace captures.

Seen correctly, Laplace's law is not causal but diagnostic. It tells us whether a boundary is aligned. If the observed curvature and pressure difference obey the relation, the boundary is coherent. If they do not, the boundary is in transition or about to fail.

This also explains why the law feels force-like. After alignment, the system behaves *as if* there were a pulling tension. But that "force" is not doing work; it is the shadow of alignment already achieved. Using it as a driver is like mistaking a balance sheet for an engine.

Placed correctly, Laplace's law becomes more powerful, not less. It stops pretending to explain formation and instead becomes a precise statement of equilibrium conditions after resolution. It belongs alongside Bernoulli's principle and conservation laws—not as origins of behavior, but as summaries of what holds once coherence is established.

The bubble does not form because Laplace's law demands it. Laplace's law holds because the bubble has already formed coherently.

With this correction, the story finally flows in the right direction:

delta appears → alignment occurs → boundary stabilizes → geometry settles → Laplace relation applies.

Nothing mystical. Nothing forced. Just consequence.

With Laplace placed where it belongs, the final picture of bubble formation becomes clear—not as construction, not as optimization, but as resolution completing itself into a temporary, coherent structure.

That is where we now turn.

8.5 — Bubble Formation as Consequential Resolution

A bubble is not built. It is not assembled by forces, nor sculpted by intent. It appears when a specific sequence of conditions is satisfied.

That appearance is not a goal achieved, but a consequence of resolution completing itself.

The sequence is always the same, regardless of scale or context.

First, a pressure delta is introduced—gently, within Alignment Threshold Time. This delta is small, local, and non-disruptive. It does not demand motion; it creates the *possibility* of reorganization.

Second, a boundary capable of mediating that delta must exist. The soap film does not arise because air pushes liquid outward. It arises because surfactant molecules organize into an alignment boundary that can host the misalignment between inside and outside air. This organization happens only if timing allows ripple alignment to propagate across the interface.

Third, once the boundary exists, it must settle into a geometry that does not introduce additional deltas along itself. The sphere emerges—not as a preference, but as the only equipotential solution that allows the boundary to remain coherent.

At no point does the system aim to "make a bubble." The bubble is what remains when nothing further needs to resolve.

This is why bubble size depends so sensitively on the solution. Concentration, viscosity, evaporation rate, and contamination all affect alignment capacity and ATT. A solution that can sustain lateral ripple coherence allows large bubbles. One that cannot produces only small bubbles or none at all. The volume of solution matters far less than its alignment quality.

This also explains why bubbles often wobble, oscillate, or briefly deform during formation. These are not instabilities; they are alignment corrections. Ripple fields are redistributing, thickness is equalizing, curvature is settling. Once alignment completes, the shape stabilizes. If alignment cannot complete, the bubble never truly forms—it collapses mid-process.

Rupture follows the same logic in reverse.

As external conditions change—evaporation thins the film, vibration introduces rapid deltas, airflow increases gradients—the alignment boundary is asked to mediate more misalignment than it can sustain. Ripple coherence drops below threshold. The boundary does not "tear"; it ceases to function as a boundary. The bubble disappears, often fragmenting into smaller droplets that each represent residual, locally aligned regions.

In this sense, droplets are not debris. They are leftover coherence.

Bubble formation and rupture are thus symmetric processes. One absorbs energy as alignment work is done to create the boundary. The other releases energy as that alignment collapses and redistributes. No hidden exothermic mystery exists—only field reconfiguration.

Seen through DELTA, bubbles are among the most honest physical phenomena. They expose, without disguise, how misalignment resolves into temporary structure when conditions allow—and vanishes the moment they do not.

The bubble is not delicate. It is precise.

Understanding bubble formation as consequential resolution completes the boundary story. What remains is to address a subtle but important question: if ripple tension is the true primitive, why does the language of surface tension persist—and when is it still safe to use it?

That clarification comes next.

8.6 — Energy Exchange During Formation and Rupture

Bubble formation and rupture are often described energetically in vague or even contradictory terms—formation feels like "work being done," rupture feels like a sudden "release." These impressions are correct, but they are usually explained with the wrong mechanism.

Nothing is being stretched like a spring, and nothing snaps back elastically. What is exchanged is alignment energy, not stored force.

When a bubble forms, energy is absorbed, but not in the way a material absorbs strain energy. The work done during formation goes into creating and maintaining lateral alignment across the boundary. Surfactant molecules migrate, orient, and couple. Thickness redistributes. Ripple coherence establishes itself over a growing surface. All of this requires energy because it moves the system away from random molecular arrangements into a highly organized, conditional state.

This energy is not stored at a point. It is distributed across the alignment field.

That is why bubble formation feels gentle but effortful. The energy input is small per unit area, but it must be supplied continuously and coherently as the boundary expands. If energy input is too abrupt, it does not help—it disrupts alignment and prevents formation altogether.

Rupture reverses this process.

When ripple coherence collapses—due to thinning, contamination, vibration, or rapid pressure change—the alignment field can no longer be sustained. The organized boundary dissolves into a less constrained configuration. The energy that was previously tied up in maintaining lateral coherence is released back into the surrounding field.

This release is not explosive by default. It appears dramatic only because alignment collapses faster than it was built. The system moves from a highly constrained, coherent state to a more distributed one in a short time. The visible effects—snap, shimmer, fine droplets—are simply the field redistributing alignment energy across available pathways.

Droplets formed during rupture are not evidence of stored elastic energy. They are evidence of local alignment surviving global collapse. Each droplet represents a small region where coherence momentarily persists, carrying a fraction of the alignment energy forward in a new configuration.

This explains why rupture can appear exothermic without violating any principle. No energy is created. Energy that was previously doing quiet organizational work is now free to manifest as motion, vibration, and heat. Alignment energy becomes kinetic and thermal energy as the boundary ceases to exist.

In DELTA terms:

- Formation absorbs energy because alignment work is being done against disorder.
- Rupture releases energy because alignment constraints are being removed.

There is no mystery and no special bubble physics here. This is the same exchange that occurs whenever structure forms and dissolves—crystallization, fracture, relaxation, melting. Bubbles simply make the process visible.

The key insight is this: energy exchange in bubbles is not about stretching and snapping. It is about building and dissolving alignment.

Once this is understood, the last remaining question becomes almost historical. If ripple tension explains boundary behavior so cleanly, why has "surface tension" survived as the dominant term for so long—and when, if ever, is it still acceptable to use it?

That clarification closes the chapter.

8.7 — Why "Surface Tension" Persists as a Term

The persistence of the term *surface tension* is not a failure of physics. It is a success of mathematics paired with an accident of language.

Historically, surface phenomena were approached from the outside in. What could be measured were forces, pressures, and shapes—not molecular organization or field alignment. When curved interfaces behaved *as if* they were under tension, the language of elasticity offered a convenient bridge. It allowed equations to be written, predictions to be made, and experiments to match calculation.

The equations worked. The metaphor quietly overstayed its usefulness.

Surface tension survived because it is an effective description after alignment has already occurred. Once a boundary is coherent, once ripple alignment is complete, the interface responds predictably to small perturbations. In that narrow regime, the boundary behaves mathematically *like* a stretched surface. Using a tension term becomes acceptable shorthand.

But shorthand hardened into explanation.

Over time, the force metaphor slipped upstream in the causal chain. Surface tension was no longer treated as a consequence of boundary alignment, but as its cause. This reversal introduced conceptual friction that never quite resolved—why surfaces "want" to shrink, why tiny amounts of soap change everything, why rupture is sudden rather than elastic.

Ripple tension resolves these puzzles by restoring sequence.

Alignment comes first. Effective tension appears second.

Surface tension is what ripple tension looks like after coherence is achieved and frozen into description. It is a macroscopic shadow cast by lateral alignment at the boundary. Using it is not wrong—but using it as a driver is.

This distinction matters because surface tension language fails precisely where understanding is most needed: during formation, transition, and breakdown. It explains stable bubbles but not their birth. It predicts pressure differences but not why boundaries appear at all. It accounts for curvature but not why only certain solutions work.

Ripple tension, by contrast, explains existence, persistence, and failure in one framework.

So when is surface tension still safe to use?

- When the boundary is already aligned
- When changes are slow and small
- When the interface remains coherent
- When equations are being used descriptively, not causally

Outside these conditions, surface tension becomes misleading. It hides timing, alignment, and capacity behind a force metaphor that cannot explain sudden collapse or selective formation.

This is not a call to erase classical terminology. It is a call to demote it—from cause to consequence, from primitive to proxy.

Physics often advances this way. Concepts begin as metaphors, mature into equations, and are later replaced as primitives by deeper structure. Force itself underwent this transition. Surface tension is ready for the same correction.

Ripple tension is not a new force. It is the reason any force-like description works at all.

With that clarification, the boundary story is complete. Pressure introduces delta. Alignment boundaries attempt to hold it. Ripple tension determines whether they succeed. When they fail, resolution does not stop—it moves inward.

That inward resolution has a name.

It is called friction.

And it does not resist motion. It resolves delta.

From Boundaries to Bulk Resolution

This chapter completes the story of boundaries—not as stretched skins or force-bearing surfaces, but as alignment structures that temporarily hold misalignment between fields. Once that shift is made, everything about bubbles becomes straightforward.

Soap films exist because alignment is possible. Ripple tension holds because lateral coherence is maintained. Spheres appear because equipotential boundaries are required. Laplace's law works because alignment has already completed. Rupture occurs because alignment can no longer be sustained.

Nothing pulls. Nothing prefers. Nothing snaps.

Boundaries do one thing only: they delay resolution by organizing it.

As long as that organization holds, delta remains localized and coherent. The moment it fails, resolution does not disappear—it spreads. What was once held at a surface diffuses into the volume. The problem does not end at the boundary; it moves inward.

This is the crucial handoff.

When a boundary can no longer mediate delta, the system must resolve misalignment through its bulk. That bulk resolution appears to us as resistance, slowing, heating, drag, or dissipation. We give it different names in solids and fluids, but the mechanism is the same.

It is not opposition. It is redistribution.

The next chapter makes this explicit by dismantling one of the most entrenched distinctions in physics: solid friction versus fluid friction. Under DELTA, both are expressions of the same process—delta being resolved through internal alignment pathways when surface mediation fails or is insufficient.

Static friction, kinetic friction, viscous drag, terminal velocity—all of them will be shown to follow from the same logic already established: resolution proceeding at the rate alignment allows.

Once boundaries are understood, friction stops being mysterious. It stops resisting motion. It starts explaining it.

We now leave the surface and enter the bulk.

Where alignment does not hold, delta resolves anyway.

A bubble is not pulled into shape. It settles into the only boundary that can hold its delta coherently.

Chapter 9 — Friction Is Delta Resolution

Why Friction Has Always Felt Like a Nuisance

Friction is almost always introduced as the villain of motion. It is the reason machines lose efficiency, the reason objects slow down, the reason energy is "wasted" as heat. In problem sets, friction is the term that ruins elegance. Remove it, and equations become clean. Add it back, and everything becomes messy, empirical, and inconvenient.

This discomfort is not accidental. It comes from framing friction as opposition.

We are taught to imagine motion as something that should continue indefinitely unless an opposing force intervenes. In that picture, friction appears as an external enemy—something that fights motion, drains energy, and degrades performance. The metaphor feels intuitive, but it has a subtle flaw: nothing in nature is trying to preserve motion.

Motion persists only when alignment allows it.

This is why friction feels so different in solids and fluids. In solids, it is blamed on roughness and contact. In fluids, it is blamed on viscosity and drag. Two separate explanations are offered, two separate formulas are memorized, and the deeper question is quietly avoided: *why should motion decay at all?*

The usual answer—"because friction acts"—simply repeats the problem in different words.

From a DELTA perspective, friction is not an intruder. It is what happens when misalignment spreads into the system instead of being held at a boundary. It is not resistance to motion; it is the completion of resolution that motion initiated.

When an object begins to move, it introduces delta—between layers, between structures, between the object and its environment. If that delta can be held coherently, motion persists smoothly. If it cannot, resolution moves inward. Internal reorganization begins. Energy redistributes across degrees of freedom. We experience this redistribution as friction.

This is why friction is unavoidable. It is not an add-on. It is the system finishing what motion started.

Once this is seen, friction stops being a nuisance and starts being informative. It tells us where alignment failed, how delta is being absorbed, and why motion cannot remain coherent under certain conditions. It also explains why removing friction entirely produces unphysical fantasies—systems with no way to resolve misalignment.

This chapter repositions friction where it belongs: not as resistance, but as delta resolution through internal alignment diffusion. Solids and fluids will be shown to obey the same logic. Static and kinetic friction will fall into place. Terminal velocity will stop feeling like a balance trick.

Friction has never been the problem. It has been the explanation.

Once friction is understood correctly, motion becomes honest again—and loss stops pretending to be mysterious.

9.1 — Friction Is Not Opposition: It Is Resolution

The deepest mistake in how friction is taught is the assumption that motion is the natural state and friction is what interrupts it. In reality, motion is conditional. It persists only as long as misalignment can be carried coherently without leaking into the system. Friction appears precisely at the moment that coherence fails.

Nothing is being opposed.

When an object moves—across a surface, through a fluid, or within a medium—it introduces delta. Layers shear. Contact points misalign. Velocity gradients form. As long as these deltas can be held at interfaces or distributed smoothly, motion continues with little penalty. When they cannot, resolution shifts inward. The system begins to reorganize internally.

That internal reorganization is what we call friction.

Friction does not act *against* motion. It acts on misalignment created by motion. It is the mechanism by which excess delta is absorbed into internal degrees of freedom—vibrations, molecular rearrangements, micro-deformations, and ultimately heat. Motion slows not because something is pushing back, but because the system is busy resolving what motion disturbed.

This is why friction always appears where alignment pathways are limited.

In a perfectly aligned system, motion would require no correction. But such systems do not exist outside idealization. Real materials have structure. Real fluids have viscosity. Real environments impose constraints. Each of these limits how much delta can be carried without internal adjustment.

Seen this way, friction is not an external force added to equations. It is the inevitable completion of resolution when delta exceeds what can be held at boundaries.

This reframing explains several long-standing puzzles at once.

First, it explains why friction depends on *how* motion occurs, not just *that* it occurs. Gentle motion produces little friction because delta is introduced slowly and remains within alignment capacity. Abrupt motion produces more friction because delta arrives faster than it can be distributed.

Second, it explains why friction generates heat. Heat is not "lost energy." It is energy that has been demoted from coherent motion into distributed alignment work. Order has not been destroyed; it has been spread across many degrees of freedom.

Third, it explains why friction is unavoidable. Any system that allows motion must also allow resolution. Preventing friction entirely would require preventing internal reorganization—which would mean preventing motion itself.

In DELTA terms, friction is not a failure mode. It is a necessary exit channel for misalignment.

Once friction is seen this way, the artificial divide between solid friction and fluid friction begins to dissolve. Both are simply different geometries of the same process: delta entering a system faster than it can remain coherently expressed as motion.

That unification is the next step.

9.2 — Unifying Solid and Fluid Friction

The traditional separation between solid friction and fluid friction is largely historical. It arose from surface-level differences in how motion looks, not from differences in how misalignment is resolved. Once friction is understood as delta resolution, the distinction collapses naturally.

In solids, friction is said to arise from roughness, interlocking asperities, and normal forces. In fluids, it is attributed to viscosity, drag, and velocity gradients. These descriptions are useful for calculation, but they obscure a deeper unity: in both cases, motion introduces relative misalignment between adjacent layers, and friction is how that misalignment is absorbed.

The difference is geometry, not mechanism.

In a solid, misalignment is concentrated at contact points. Motion causes micro-regions to deform, detach, reattach, and rearrange. These processes are not smooth. They occur in bursts as local alignment thresholds are crossed. Energy is redistributed into lattice vibrations, micro-fractures, and internal stress relaxation. We experience this as solid friction.

In a fluid, misalignment is distributed continuously. Motion creates velocity gradients between neighboring layers. Each layer tries to remain aligned with its neighbors, but viscosity limits how quickly this can happen. Delta spreads gradually through the medium. Energy is absorbed as molecular reorganization and thermal motion. We experience this as drag.

In both cases, the system is not resisting motion. It is resolving the misalignment motion creates.

This explains why increasing surface smoothness in solids reduces friction only up to a point. Once asperities are minimized, friction persists because misalignment must still be resolved through internal deformation. Similarly, streamlining objects in fluids reduces drag, but never eliminates it. Alignment capacity improves, but it does not become infinite.

This also explains why lubrication works in both regimes. A lubricant does not remove friction by making things slippery; it changes the alignment pathway. In solids, it replaces abrupt contact misalignment with smoother, distributed shear. In fluids, it alters viscosity and boundary layers. In both cases, it spreads delta over space and time, keeping resolution within ATT.

The same principle governs granular materials, gels, foams, and biological tissues. They occupy intermediate geometries, but the mechanism is unchanged. Wherever layers move relative to one another, delta appears. Wherever delta exceeds alignment capacity, resolution follows.

Friction, then, is not a property of materials. It is a property of motion interacting with structure.

Solids and fluids merely present different structural constraints on how alignment can propagate. One resolves delta through localized deformation. The other through distributed shear. Both obey the same DELTA logic.

With this unification in place, a classic puzzle becomes easier to resolve: why static friction behaves so differently from kinetic friction—even though motion is involved in both.

That inversion is where friction's real character is revealed.

9.3 — Static vs Kinetic Friction Reinterpreted

Static friction has always been taught as a mystery of stubbornness: objects "refuse" to move until a threshold force is exceeded. Kinetic friction, by contrast, appears calmer and smaller, as if motion somehow makes resistance easier. In force language, this is awkward. Why should resistance be *greater* when nothing is moving?

Under DELTA, the paradox disappears.

Static friction is not resistance to motion. It is alignment preservation.

When two surfaces are in contact and at rest relative to each other, they are not passive. They are actively aligned. Contact points—whether microscopic asperities in solids or structured boundary layers in soft materials—have already resolved their mutual delta. The system has settled into a locally coherent configuration. As long as that coherence is maintained, motion is unnecessary.

To initiate motion, this existing alignment must be broken.

That breaking is not free. It requires introducing delta faster than the contact region's Alignment Threshold Time allows. Until that threshold is crossed, any applied influence is quietly absorbed as elastic deformation, micro-adjustment, or stress redistribution. Nothing moves because the system is still capable of preserving alignment internally.

Static friction, then, measures how much delta the existing alignment can absorb without reorganizing.

The moment motion begins, the situation changes fundamentally.

Once relative motion starts, the system is no longer trying to preserve a settled alignment. It has accepted reorganization. Contact points are continually forming and dissolving. Misalignment is no longer resisted; it is being resolved continuously. The system shifts from alignment preservation to alignment diffusion.

This is kinetic friction.

Kinetic friction is lower not because the system is helping motion, but because the costly act—breaking coherence—has already occurred. Resolution is now flowing rather than being blocked. Delta introduced by motion is spread out over time instead of being absorbed elastically at fixed contact points.

This explains several long-standing observations:

- Static friction depends strongly on contact history because it reflects how well alignment has settled.
- Vibrations reduce static friction because they inject small deltas that partially erode coherence.
- Once motion begins, friction stabilizes because the system has entered a steady regime of delta diffusion.
- Smooth motion feels easier not because resistance vanished, but because alignment is no longer being defended.

The static–kinetic transition is therefore not about overcoming a larger opposing force. It is about crossing an alignment threshold. Below that threshold, the system prefers coherence. Above it, the system reorganizes and allows resolution to proceed dynamically.

In DELTA terms:

- Static friction is the cost of *breaking alignment*.
- Kinetic friction is the cost of *maintaining resolution*.

This also clarifies why sudden jerks succeed where slow pushes fail. A rapid change violates ATT and forces reorganization. A slow push gives the system time to absorb delta elastically and remain aligned.

Once this distinction is clear, friction in fluids becomes easier to understand as well—because there, alignment is never fully preserved. Resolution begins immediately, smoothly, and predictably.

That brings us to one of the cleanest demonstrations of friction as delta resolution: Stokes law and terminal velocity.

9.4 — Stokes Law and Terminal Velocity

Terminal velocity is usually introduced as a balance trick: gravity pulls down, drag pulls up, forces cancel, acceleration stops. The math works. The explanation does not. It leaves a lingering question—*why does the system settle instead of oscillating or continuing to accelerate?* What exactly is being "balanced"?

Under DELTA, terminal velocity is not a force stalemate. It is a resolution ceiling.

When an object is released into a fluid, the initial state carries a large delta. There is a sharp misalignment between the object's motion and the surrounding medium, which is initially at rest. Alignment pathways are undeveloped. The object accelerates rapidly because the field has not yet reorganized to absorb the introduced delta.

This early acceleration is not caused by an unopposed force. It is caused by insufficient alignment.

As the object speeds up, it creates velocity gradients in the surrounding fluid. Layers near the object are dragged along, alignment spreads outward, and internal reorganization begins. The fluid starts doing alignment work—molecular rearrangement, shear diffusion, and thermalization. This is drag, but not as opposition. It is delta being resolved into the fluid.

Stokes law captures this regime precisely because it applies where alignment diffusion is smooth and proportional. Drag increases linearly with velocity because the rate of delta introduction increases linearly. The faster the object moves, the faster misalignment is injected—and the faster the fluid resolves it.

Terminal velocity appears when these two rates match.

At that point, the system has reached a dynamic equilibrium: the rate at which motion introduces delta equals the rate at which the fluid can absorb it through alignment diffusion. There is no further buildup of unresolved misalignment, so there is no reason for acceleration to continue.

Acceleration stops not because forces cancel, but because resolution capacity is saturated.

This explains why terminal velocity is stable. Small deviations decay automatically. If the object speeds up slightly, delta arrives faster than it can be resolved, increasing drag and restoring the previous rate. If it slows down, alignment diffusion briefly exceeds delta injection, reducing resistance and allowing speed to recover.

No force coordination is required. The field self-regulates.

This framing also explains why terminal velocity depends on fluid viscosity, object size, and shape. These factors do not change "opposing force strength." They change alignment bandwidth—how quickly the surrounding medium can reorganize to absorb misalignment.

In low-viscosity fluids, alignment spreads slowly, so terminal velocity is high. In high-viscosity fluids, alignment spreads quickly, so terminal velocity is low. The governing variable is not force magnitude but delta resolution rate.

Seen this way, Stokes law is not a special case. It is a clean demonstration of friction as delta resolution operating in its most transparent regime—smooth, continuous, and time-governed.

Terminal velocity is not where motion is stopped. It is where motion becomes fully accounted for.

With this understanding, dissipation stops looking like loss and starts looking like distribution. The next section completes the picture by explaining where the resolved delta actually goes—and why "heat" is just alignment diffused beyond perception.

9.5 — Dissipation as Alignment Diffusion

Dissipation is usually described with a sigh. Energy is said to be "lost," motion "dies out," and heat appears as an afterthought—useful for engines, inconvenient for everything else. This language subtly implies failure, as if nature mismanages energy once friction enters the scene.

Under DELTA, dissipation is neither loss nor failure. It is alignment diffusion.

When motion introduces delta and that delta can no longer be carried coherently as organized movement, it does not vanish. It spreads. The system redistributes misalignment across increasing numbers of internal degrees of freedom—molecular vibrations, rotations, lattice oscillations, microscopic flow patterns. What we call heat is simply alignment work distributed so widely that coherence is no longer visible.

Nothing disappears. Nothing is wasted.

Order is not destroyed; it is diluted.

This is why dissipation is irreversible in practice. Once alignment has diffused across millions of microstates, reassembling it into a single, coherent motion would require an external intervention that reintroduces delta in a controlled way. Nature does not prohibit reversal; it merely makes it statistically implausible.

This also clarifies the relationship between dissipation and entropy. Entropy increases not because energy degrades, but because alignment becomes less concentrated. Coherent motion represents alignment held along a narrow channel. Dissipation represents alignment distributed across many channels. The total energy remains the same; its organizational density changes.

Seen this way, frictional heating is not a side effect—it is the destination. Motion is a temporary carrier of alignment. When that carrier exceeds its capacity, alignment migrates into thermal degrees of freedom, where it is stored as microscopic motion rather than macroscopic direction.

This perspective resolves a common confusion: why dissipation increases with speed. Faster motion introduces delta at a higher rate. Alignment must therefore diffuse more rapidly. The result is greater heating—not because friction "pushes harder," but because more misalignment is being resolved per unit time.

It also explains why materials differ so dramatically in dissipation behavior. Crystalline solids, amorphous solids, viscous fluids, gases—all offer different alignment pathways. Some diffuse alignment smoothly and quietly. Others concentrate it into stress, wear, or damage before releasing it. These differences are not about resisting motion; they are about how alignment is allowed to spread.

In DELTA terms, dissipation is the system completing the story motion began. Motion concentrates alignment. Friction releases it. Heat is the signature of alignment that has become fully shared.

There is no tragic ending here. Only redistribution.

With dissipation understood as alignment diffusion, the last missing piece becomes timing. Why gentle motion survives longer, why abrupt motion damages systems, and why rate matters more than magnitude—all of it returns us to the same governing constraint.

That constraint is Alignment Threshold Time.

And that is where we go next.

9.6 — Rate Matters More Than Magnitude

One of the most persistent misconceptions in physics is that outcomes are governed primarily by *how much*—how much force, how much pressure, how much energy. In frictional systems, this belief repeatedly fails. Gentle actions succeed where powerful ones fail. Slow motions preserve structure where fast motions destroy it.

The reason is simple: alignment has a speed limit.

Every material, every fluid, every interface has a finite rate at which it can reorganize internally to accommodate misalignment. This rate is its Alignment Threshold Time. When delta is introduced slowly enough, alignment pathways open, redistribute stress, and diffuse misalignment without damage. When delta is introduced faster than alignment can propagate, coherence breaks.

Magnitude is tolerated. Speed is not.

This is why polishing works while grinding damages. Polishing introduces small deltas repeatedly within ATT, allowing alignment to diffuse smoothly. Grinding injects delta faster than alignment can spread, concentrating stress until material structure fails. The difference is not energy input, but rate of misalignment delivery.

The same principle governs lubrication. A lubricant does not eliminate friction by reducing force. It increases alignment bandwidth. It spreads shear over time and space, lowering the effective rate at which delta is imposed on contact regions. Friction becomes gentler not because less energy is involved, but because alignment is allowed to keep up.

This also explains why sudden jerks overcome static friction while steady pushes fail. A slow push gives the system time to absorb delta elastically and remain aligned. A sharp impulse violates ATT, forcing reorganization. Motion begins not because force increased, but because timing crossed a threshold.

In fluids, the same logic applies. Laminar flow persists when velocity gradients change slowly. Turbulence erupts when gradients evolve faster than alignment can diffuse. Viscosity is not resistance; it is a measure of how quickly alignment can spread.

Rate also governs heating and wear. Slow motion dissipates alignment quietly into thermal motion. Rapid motion concentrates misalignment into stress, micro-fractures, and damage before it can diffuse. This is why high-speed friction is destructive and low-speed friction is survivable.

In DELTA terms, systems fail not from excess energy, but from excessive resolution demand per unit time.

Once rate is recognized as the controlling variable, friction stops being unpredictable. It becomes a timing problem. And timing, unlike force magnitude, is something systems can often be designed to respect.

But even alignment has limits. When delta is repeatedly imposed faster than resolution allows, damage accumulates. Alignment pathways degrade. Friction ceases to be protective and becomes destructive.

That transition is the final topic of this chapter.

9.7 — When Friction Becomes Destructive

Friction is protective only as long as alignment pathways remain intact. When delta is imposed repeatedly—or abruptly—beyond what those pathways can handle, friction changes character. It stops diffusing misalignment and starts recording damage.

This transition is not sudden in cause, even if it appears sudden in effect.

Under normal conditions, friction spreads delta into internal degrees of freedom: molecular rearrangement, elastic deformation, thermal motion. Alignment absorbs stress and releases it harmlessly. But every material has a finite capacity for such accommodation.

When resolution is demanded faster than alignment can recover—again and again—the system begins to accumulate unresolved micro-misalignment.

That accumulation is damage.

In solids, this appears as wear, fatigue, micro-cracking, and plastic deformation. Contact regions that once reorganized elastically now deform irreversibly. Alignment pathways become scarred. Each subsequent motion finds fewer routes to diffuse delta, forcing stress into ever-smaller regions. Friction increases, heating rises, and failure accelerates—not because friction "grew stronger," but because alignment capacity degraded.

In fluids, destructive friction appears as turbulence, erosion, cavitation, and shear breakdown. When velocity gradients exceed ATT repeatedly, coherent flow fragments into chaotic eddies. Energy that once diffused smoothly now concentrates into vortices, pressure spikes, and localized collapse. The fluid is not misbehaving; it is failing to keep up.

This is why damage is cumulative. Alignment pathways do not reset instantly. Once degraded, they reduce future resolution bandwidth. A system that was once tolerant becomes fragile. The same delta that was previously diffused safely now produces rapid failure.

Friction, then, is a guardian until it is overworked.

This reframing explains why rest matters, why cooling restores performance, and why materials "heal" when given time. Recovery is not mystical. It is the slow re-establishment of alignment pathways—annealing in solids, relaxation in fluids, redistribution in interfaces. Given time within ATT, coherence can return.

It also explains why high-speed systems demand special design. Bearings, lubricants, surface treatments, and flow control do not remove friction. They protect alignment capacity by spreading delta across space and time, keeping resolution demands below destructive thresholds.

In DELTA terms, destruction occurs when friction is asked to do the impossible: resolve misalignment faster than alignment can exist.

Friction does not fail. Alignment does.

With this, the story of friction reaches its natural conclusion. Delta introduced by motion must resolve somewhere—at boundaries, within bulk, or across time. When resolution remains coherent, systems flow smoothly. When it does not, chaos and damage appear.

The final step is to understand how coherence can persist *despite* motion—how some systems conserve alignment instead of dissipating it.

That brings us to the next chapter, where friction hands the baton to flow itself, and Bernoulli's principle is finally read as a statement about where alignment survives.

From Friction to Flow Coherence

This chapter closes the loop on one of physics' most misunderstood ideas. Friction is no longer an enemy of motion, no longer a tax paid to reality, and no longer a mysterious sink where energy disappears. It is where delta goes when it cannot remain coherent.

When alignment pathways are available, friction diffuses misalignment quietly. When they are overloaded, friction records damage. When they collapse entirely, chaos follows.

In every case, the role of friction is the same: to complete resolution.

This insight dissolves artificial divisions. Solids and fluids differ only in geometry. Static and kinetic friction differ only in timing. Dissipation differs from motion only in how widely alignment is spread. Once these distinctions are reframed, friction becomes predictable—not through force balances, but through resolution capacity.

But this also reveals a complementary truth.

Not all motion dissipates immediately.

Some systems manage to carry delta without diffusing it. They redirect alignment instead of absorbing it. They allow motion to persist not by fighting friction, but by staying within coherence bandwidth. In such systems, energy does not vanish into heat; it travels.

That travel is what classical physics tried to capture under conservation principles—often without explaining *how* coherence survived along the way. Bernoulli's principle is one of those attempts. It is usually presented as an energy accounting rule. In reality, it is a statement about where alignment remains intact and where it leaks.

To understand flow honestly, friction must now step aside and let coherence speak.

The next chapter does not deny dissipation. It shows the narrow corridor where dissipation is delayed.

Where alignment holds, flow persists. Where it fails, friction resumes.

We now follow alignment itself— into Bernoulli revisited.

Friction does not fight motion. It completes the resolution motion began.

Chapter 10 — Bernoulli Revisited

The Equation That Worked Without Explaining Why

Bernoulli's equation is one of those rare formulas that almost everyone uses and very few trust.

It appears early in education as a neat balance between pressure, velocity, and height. Students learn to move terms around, plug in numbers, and get answers that match experiments—until they don't. At that point, "losses" are added. Coefficients appear. The elegance fades, and Bernoulli becomes conditional, approximate, and slightly embarrassing.

And yet, something about it refuses to die.

Engineers rely on it. Nature seems to respect it. Blood flows, air lifts wings, rivers accelerate through constrictions. The equation works often enough that abandoning it is impossible—but explaining *why* it works remains oddly unsatisfying.

The discomfort comes from treating Bernoulli as an energy accounting identity.

When framed this way, the equation feels like bookkeeping without a mechanism. Pressure becomes stored energy, velocity becomes kinetic energy, height becomes potential energy, and their sum remains constant—unless friction "steals" some of it. This story is tidy, but it explains nothing about *how* the system moves energy between these forms, or why it does so smoothly in some cases and catastrophically in others.

The real mystery is not conservation. It is coherence.

Why does energy remain organized along a streamline instead of diffusing immediately into heat? Why does acceleration happen quietly without apparent force input? Why does a slight roughness destroy Bernoulli behavior while enormous pressure differences do not?

Bernoulli's equation does not answer these questions because it was never meant to. It records a pattern that emerges only when alignment survives. It is not the cause of flow behavior; it is the signature of a special regime where resolution remains coherent.

Seen this way, Bernoulli stops being a law about energy and becomes a diagnostic of alignment.

When pressure, velocity, and elevation exchange smoothly, alignment is intact. When the sum fails to hold, alignment has leaked. The equation itself is silent on why this happens—but it tells us *where* it happens.

This chapter revisits Bernoulli with that correction in mind. Nothing will be rejected. No mathematics will be denied. But the causal chain will be restored.

Bernoulli does not govern flow. Flow governs when Bernoulli applies.

Once this is understood, "losses" stop being mysterious, and acceleration stops feeling magical. The equation finally says what it has been saying all along—quietly, faithfully, and without explanation.

It tells us where coherence survives.

The sections that follow unpack this, step by step, until Bernoulli emerges not as a fragile formula, but as one of the cleanest windows into DELTA-controlled flow.

10.1 — What Bernoulli Really Conserves

Bernoulli's equation is usually introduced as a statement of energy conservation along a streamline. Pressure energy, kinetic energy, and gravitational potential energy are said to trade places while their sum remains constant. This is mathematically correct—but physically incomplete.

Energy is always conserved. Bernoulli applies only sometimes.

That fact alone tells us energy cannot be the conserved *quantity* Bernoulli is really tracking. If energy conservation were sufficient, Bernoulli would never fail. Yet it fails routinely—in turbulence, in sharp bends, in rough pipes, in sudden expansions—precisely where energy is still conserved but something else has changed.

What Bernoulli actually conserves is aligned resolution capacity.

Along a streamline where Bernoulli holds, delta is being resolved without spilling into internal degrees of freedom. Misalignment introduced by motion is continuously redistributed between pressure, velocity, and elevation without breaking coherence. The system remains organized. Nothing leaks inward. Nothing diffuses prematurely.

Pressure, velocity, and height are not "forms of energy" in this context. They are expressions of where alignment currently resides.

High pressure corresponds to alignment stored as compression—delta held quietly in the field. High velocity corresponds to alignment expressed as organized motion—delta carried forward coherently. Height corresponds to alignment stored against a large-scale field gradient. Bernoulli's sum remains constant only as long as alignment can move freely between these expressions without fragmenting.

This reframing explains why Bernoulli is conditional by nature.

It applies only when:

- flow is smooth enough for alignment to propagate,
- gradients change slowly enough to respect ATT,
- boundaries do not force premature resolution,
- internal reorganization is minimal.

In these conditions, alignment is conserved even as its *form* changes.

The moment alignment can no longer be carried coherently—when velocity gradients steepen too quickly, when boundaries disrupt lateral coupling, when surface roughness injects rapid delta—alignment diffuses. Energy is still conserved, but it is no longer organized. Bernoulli fails not because energy disappeared, but because coherence did.

This also clarifies why Bernoulli is silent about time. It assumes that alignment redistribution happens within the system's temporal bandwidth. The equation has no explicit clock because it presupposes timing compatibility. When timing fails, the equation collapses without warning.

Seen correctly, Bernoulli is not a universal conservation law.

It is a coherence certificate.

If pressure, velocity, and elevation exchange smoothly, alignment survives. If they do not, alignment has leaked—into heat, turbulence, sound, or internal stress. The "loss terms" added later are not corrections to energy conservation; they are markers of where alignment escaped the Bernoulli regime.

Bernoulli does not promise that flow will behave. It tells us when it *can*.

With this understanding in place, the next step becomes clear. If Bernoulli tracks alignment rather than energy, then the familiar pressure–velocity trade-off is not a conversion of energy types. It is something subtler: an exchange of alignment expression.

That exchange is where flow actually lives.

10.2 — Alignment Exchange, Not Energy Bookkeeping

The most familiar move in Bernoulli's equation is the apparent trade: pressure goes down, velocity goes up; velocity goes down, pressure recovers. This is usually narrated as energy changing forms—pressure energy converting into kinetic energy and back again. The arithmetic works, but the story misleads.

Nothing is being converted. Nothing is being spent.

What is happening is an exchange of alignment expression.

When flow accelerates through a constriction, the system is not gaining kinetic energy by doing work. It is re-expressing the same alignment in a different, more directional way. Pressure decreases not because energy has been lost, but because alignment is no longer being held as isotropic compression. It is being carried forward as organized motion.

Pressure and velocity are not rivals. They are two ways alignment can exist.

Pressure is alignment held quietly—distributed, omnidirectional, locally stable. Velocity is alignment made directional—channeled along a path, extended in space. When a flow enters a region that geometrically favors direction (a narrowing, a smooth curve, a downhill path), alignment naturally migrates toward velocity. When the geometry relaxes, alignment thickens back into pressure.

No force performs this exchange. No agent decides it.

The field simply follows the path that allows alignment to remain coherent.

This reframing dissolves a long-standing confusion: why acceleration in fluids often occurs without any identifiable pushing force. The fluid does not need to be pushed into motion. It is already aligned. The geometry merely permits alignment to reorganize.

Seen this way, a nozzle does not "add speed." It removes places for alignment to hide.

As cross-sectional area shrinks, alignment can no longer remain isotropic. It must become directional. Velocity increases as a *geometric necessity*, not as a result of energy injection. Pressure drops because compression is no longer the dominant alignment mode.

Deceleration works the same way in reverse. In a diffuser or widening channel, alignment regains room to spread laterally. Directionality relaxes. Velocity drops. Pressure rises. Nothing is recovered or regenerated. Alignment simply thickens.

This also explains why Bernoulli flow feels fragile. Alignment exchange requires continuity. It depends on smooth boundaries, gradual transitions, and timing compatible with ATT. Sharp corners, sudden expansions, or roughness interrupt lateral coupling. Alignment cannot reorganize smoothly, so it diffuses instead of exchanging. Bernoulli fails—not because energy bookkeeping broke down, but because alignment exchange was interrupted.

The language of "loss" enters precisely here. But what is lost is not energy—it is the privilege of coherent exchange.

Bernoulli, properly understood, does not track how much energy a flow has. It tracks whether alignment can still move freely between pressure, velocity, and elevation without fragmenting.

Once that freedom is gone, the equation has nothing left to say.

The next section makes this explicit by confronting the most abused term in fluid mechanics: *losses*—and showing that they are not losses at all, but the visible footprints of alignment leaving the Bernoulli corridor.

10.3 — Why "Losses" Are Not Losses

Few words have done more quiet damage to understanding fluid mechanics than *loss*.

We speak of head loss, pressure loss, energy loss—as if something valuable has vanished somewhere along the pipe. The language suggests theft, inefficiency, or decay. It implies that Bernoulli's equation failed because friction stole energy from the system.

This is misleading in a precise way.

Nothing is lost. What is lost is coherence.

When "loss terms" appear in fluid equations, they are not accounting for missing energy. Energy is always conserved. What those terms mark is the exit of alignment from the Bernoulli regime—the moment when alignment can no longer be exchanged cleanly between pressure, velocity, and elevation.

Losses are not sinks. They are leak points.

In an ideal Bernoulli flow, alignment remains organized. Pressure thins into velocity, velocity thickens into pressure, and the system never needs to involve its internal degrees of freedom. The flow stays smooth, quiet, and predictable. As soon as conditions violate ATT—through roughness, sudden expansion, sharp curvature, or high gradients—alignment can no longer remain confined to the streamline.

It spills inward.

That spill is what we label as loss.

Alignment diffuses into shear layers, vortices, boundary-layer heating, acoustic modes, and microscopic turbulence. From the perspective of Bernoulli's equation, that alignment is gone—it can no longer participate in pressure–velocity exchange. From the perspective of the universe, nothing has disappeared. The alignment has simply changed scale and location.

This is why loss terms feel arbitrary. Engineers insert coefficients, correction factors, empirical constants—not because the physics is unclear, but because alignment leakage depends on geometry, timing, and material detail. There is no single force to compute. There is only the question: *how much alignment escaped the coherent corridor?*

Once this is seen, several confusions resolve immediately:

- A longer pipe does not "lose more energy" because it is greedy; it offers more opportunity for alignment diffusion.

- Roughness does not consume energy; it disrupts lateral coupling, forcing alignment to fragment.
- Turbulence is not violent loss; it is alignment redistributed into many interacting modes.
- Heat generation is not waste; it is alignment diluted beyond macroscopic visibility.

Losses increase when alignment exchange is interrupted. They decrease when exchange is preserved. That is why smooth transitions outperform abrupt ones, why gradual expansions matter more than absolute size, and why gentle flow regimes survive where aggressive ones fail.

In DELTA language, losses are simply where alignment exits the Bernoulli corridor.

Bernoulli does not promise lossless flow. It tells us exactly what must be preserved for lossless flow to be possible.

When that preservation fails, the equation does not break. It steps aside.

The next question, then, is not *why* Bernoulli fails—but where and when it fails, and what that failure teaches us about the limits of coherence.

That is where we turn next.

10.4 — Where Bernoulli Breaks—and Why

Bernoulli does not fail randomly. It fails precisely and instructively. Wherever it breaks, it is pointing—quietly but accurately—to a violation of coherence.

The common explanation lists conditions: turbulence, sharp bends, rough walls, sudden expansions. These are treated as exceptions, inconveniences that force us to abandon a beautiful equation. Under DELTA, they are not exceptions. They are boundary markers of where alignment can no longer be sustained.

Bernoulli holds only as long as alignment exchange can occur within Alignment Threshold Time.

When a flow is forced to turn sharply, alignment must reorganize laterally and directionally at once. If the curvature demands faster reorientation than ATT allows, alignment cannot exchange smoothly between pressure and velocity. It fragments. Vortices form. Bernoulli breaks—not because energy bookkeeping failed, but because alignment ran out of time.

Sudden expansions expose the same limit even more starkly. When flow enters a wider region abruptly, alignment loses its directional constraint instantly. Pressure recovery cannot occur smoothly because alignment has no structured pathway to thicken coherently. Instead of reorganizing, it spills into eddies and recirculation zones. The familiar "loss" appears—but what actually happened is alignment diffusion triggered by timing violation.

Rough surfaces produce a different but related failure. Microscopic protrusions inject rapid, localized deltas into the boundary layer. Alignment that would otherwise propagate smoothly is repeatedly disrupted at scales too small and too fast for recovery. Bernoulli fails locally first, then globally, as coherence erodes layer by layer.

Turbulence is not the cause of Bernoulli's breakdown. It is the symptom.

Turbulence appears when alignment exchange is repeatedly interrupted faster than it can re-establish. The flow fragments into many interacting modes because no single coherent pathway remains viable. Bernoulli's equation, which presupposes a single aligned streamline, has nothing left to describe.

This reframing explains why gradual transitions matter more than absolute conditions. A large pressure change introduced slowly preserves Bernoulli behavior. A small geometric disruption introduced abruptly destroys it. Magnitude is secondary. Rate is decisive.

It also explains why Bernoulli can hold in high-speed flows yet fail in slow, poorly shaped ones. Speed alone does not break coherence. Poor alignment pacing does.

Bernoulli breaks where:

- delta is introduced faster than alignment can reorganize,
- geometry demands simultaneous resolution in incompatible directions,
- boundary conditions inject rapid, repeated misalignment,
- internal diffusion overtakes exchange.

In every case, the equation is not wrong. It is out of jurisdiction.

Bernoulli applies only within the narrow corridor where alignment exchange remains possible. Outside that corridor, flow still obeys physics—but not Bernoulli's simplified signature.

Understanding where Bernoulli breaks turns failure into guidance. It tells us how to design flows that stay coherent longer, how to avoid premature dissipation, and how to respect the temporal limits of alignment.

That insight naturally leads to the next step: if Bernoulli marks the corridor of coherence, then engineering—and nature itself—can be seen as the art of staying inside that corridor.

That art is flow design.

10.5 — Flow Design as Coherence Management

Once Bernoulli is understood as a statement about where alignment survives, design stops being an exercise in force management and becomes an exercise in coherence management. Good flow design does not "reduce losses" by fighting friction; it paces alignment so that exchange remains possible for as long as needed.

This is why smoothness matters—not as polish, but as timing.

Gradual contractions allow alignment to thin into velocity without fragmenting. Gentle expansions allow alignment to thicken back into pressure without spilling into vortices. Long-radius bends give alignment time to reorient laterally. Each of these choices respects Alignment Threshold Time. They do not eliminate dissipation; they delay it by keeping alignment exchange viable.

In this light, streamlining is not about cheating drag. It is about protecting alignment pathways.

A streamlined shape does not push the fluid out of the way more effectively; it avoids asking the fluid to reorganize faster than it can. Sharp edges and sudden changes demand simultaneous resolution in incompatible directions. Alignment fails, diffusion begins, and Bernoulli exits. Smooth contours ask less per unit time, keeping the system within its coherence bandwidth.

This reframing also explains why scale matters in counterintuitive ways. A small, abrupt feature can destroy coherence more efficiently than a large, gentle one. Micro-roughness in the wrong place can trigger boundary-layer breakdown even when global conditions look benign. Conversely, large systems—rivers, atmospheric flows—can sustain enormous pressure differences because changes occur over long distances and times.

Nature exploits this relentlessly.

Blood vessels taper gradually, branch smoothly, and curve gently—not because biology solved equations, but because alignment must be preserved to sustain life. River deltas spread flow instead of forcing sharp turns. Lungs exchange direction and cross-section slowly to avoid coherence collapse. Wherever flow must persist, nature designs for timing, not strength.

Engineering converges on the same lesson when it works well. Diffusers outperform sudden expansions. Rounded inlets outperform sharp ones. Long nozzles outperform short throats.

These are not empirical hacks; they are expressions of alignment pacing.

Seen through DELTA, design parameters change meaning. Reynolds number, roughness, curvature, and aspect ratio are no longer abstract controls; they are proxies for alignment demand versus alignment capacity. Staying within Bernoulli's corridor becomes a matter of managing how quickly and where delta is introduced.

Flow design, then, is not optimization in the mathematical sense. It is courtesy toward alignment.

And when that courtesy is extended, Bernoulli rewards it—not by conserving energy, but by allowing coherence to travel.

The next step widens the view. If Bernoulli is really about coherence, then it should not be confined to pipes and wings. It should appear wherever alignment must move through constrained systems—across scales, across contexts.

That universality is where we go next.

10.6 — Bernoulli Across Scales

Bernoulli's principle is often taught as a narrow rule for pipes, jets, and wings—useful in engineering, distant from life. That impression is a consequence of how narrowly the equation is framed. Once Bernoulli is understood as a statement about alignment survival, its reach expands immediately.

It appears wherever alignment must travel through a constrained system without fragmenting.

At small scales, this is seen in microfluidic channels. When channels are smooth and transitions are gradual, flow remains predictable even at low Reynolds numbers. When geometry changes abruptly, coherence collapses almost instantly. The scale is tiny, the physics identical. Alignment still has an ATT. Bernoulli still holds only within it.

At biological scales, the pattern becomes unmistakable. Blood flow is not governed primarily by force or pressure magnitude, but by how gently alignment is redirected. Healthy arteries taper smoothly and curve gradually, allowing pressure and velocity to exchange without turbulence. Pathological constrictions do not fail because pressure is high, but because alignment is forced to reorganize too quickly. Bernoulli breaks there—not as a mathematical error, but as a physiological warning.

At environmental scales, rivers provide a textbook illustration. In straight, gently sloping reaches, flow accelerates and decelerates smoothly. Pressure, velocity, and elevation exchange cleanly. Bernoulli holds approximately. Where rivers encounter sharp bends, sudden drops, or artificial barriers, coherence fractures into eddies, rapids, and turbulence. Energy is not lost; alignment has escaped into chaotic modes.

At atmospheric scales, jet streams, trade winds, and convection cells all reflect the same logic. Large pressure gradients coexist with stable flow because alignment changes occur over enormous distances and times. Where those gradients steepen abruptly—storm fronts, shear layers—Bernoulli behavior collapses and turbulence dominates.

The lesson is consistent across scale:

Bernoulli does not care about size. It cares about pacing.

Scale matters only insofar as it changes alignment capacity. Larger systems often tolerate larger deltas because they also allow longer ATT. Smaller systems demand gentler transitions because alignment bandwidth is narrow. The equation itself remains agnostic. It simply reports whether coherence survived.

This also explains why Bernoulli often appears "approximately right" even in messy real-world situations. Alignment rarely collapses everywhere at once. Some regions remain coherent while others leak. Engineers and scientists intuitively average over these regimes, obtaining usable but imperfect results. The approximation works because coherence is partial, not because losses are small.

Seen this way, Bernoulli is not a fragile idealization. It is a universal coherence marker.

Wherever alignment can move without diffusing—whether in a capillary, an artery, a river, or the sky—Bernoulli-like behavior emerges. Where it cannot, no amount of energy bookkeeping will save it.

This universality hints at something deeper. If alignment governs flow in fluids, and if living systems rely on the same constraints, then Bernoulli is not just about matter in motion. It is about how organized systems survive motion without breaking.

That realization takes us beyond fluids—toward systems that breathe, circulate, and live.

10.7 — From Bernoulli to Living Systems

Once Bernoulli is understood as a statement about alignment survival, its relevance to living systems becomes unavoidable. Life does not move fluids to satisfy equations. It moves them to stay coherent under continuous motion. The same constraints that govern pipes and rivers quietly govern breathing, circulation, and flight.

Consider breathing. Air does not rush violently in and out of healthy lungs. It flows gently, guided through branching passages that taper and curve gradually. This is not inefficiency. It is alignment preservation. Rapid pressure changes would violate ATT, fragmenting flow into turbulence, increasing dissipation, and exhausting the system. Efficient breathing is not powerful breathing—it is coherent breathing.

The circulatory system tells the same story. The heart does not "push harder" to maintain flow. It sets up a rhythm that respects alignment bandwidth. Arteries are elastic not to store force, but to pace alignment exchange between pressure and velocity. Where vessels stiffen or narrow abruptly, Bernoulli-like exchange fails. Turbulence appears. Fatigue increases. Pathology begins—not because energy is missing, but because coherence is.

Flight follows the same logic. Wings do not generate lift by brute force. They shape alignment pathways so pressure can thin into velocity smoothly along one surface and thicken along another. When flow remains attached, Bernoulli behavior holds and lift persists. When alignment collapses—stall occurs—no amount of force language can explain recovery. The system has simply left the corridor of coherence.

Even at the cellular level, the pattern holds. Nutrient transport, cytoplasmic streaming, and membrane exchange all rely on gentle gradients and timed flows. Sudden forcing damages cells. Gradual exchange sustains them. Life learned, long before equations existed, that rate matters more than magnitude.

This is the deeper reason Bernoulli feels intuitively "right" in living contexts. It is not conserving energy for life's sake. It is marking where alignment remains intact long enough for function to persist. Where it holds, systems feel effortless. Where it breaks, fatigue, stress, and breakdown follow.

In DELTA terms, life survives by staying inside Bernoulli's corridor as long as possible—redirecting alignment instead of dissipating it, exchanging coherence instead of burning it.

Bernoulli, then, is not a law imposed on life. It is a constraint life learned to respect.

With this realization, this part reaches its natural close. Fluids, friction, bubbles, and flow all point to the same quiet truth: nature does not fight forces. It manages deltas through timing and alignment.

The next part will take this insight beyond physics—into systems where alignment, timing, and resolution shape not just motion, but behavior, intelligence, and survival itself.

From Flow to Universal Dynamism

With Bernoulli finally placed in its proper role, this part completes a quiet but decisive shift. Fluids, bubbles, friction, and flow no longer appear as special cases governed by clever equations. They emerge as direct expressions of how delta resolves when alignment is allowed to survive.

Pressure introduced delta. Boundaries tried to hold it. Ripple tension mediated it. Friction diffused it. Bernoulli marked where it stayed coherent.

Nothing here required force as a primitive. Nothing depended on intention, preference, or optimization. Every phenomenon followed from one rule applied honestly: delta resolves at the rate alignment permits.

Bernoulli's corridor now stands revealed—not as a conservation trick, but as a narrow passage where dynamism remains organized. Inside it, motion feels effortless. Outside it, dissipation, turbulence, and breakdown appear. The transition is not moral, mechanical, or energetic. It is temporal and structural.

This closes the chapter on "honest physics."

From here on, the same logic must be allowed to speak in domains where equations are thinner but consequences are deeper. The moment alignment, timing, and resolution govern fluids, they cannot stop governing everything else that moves, adapts, or survives.

Systems do not collapse because energy runs out. They collapse because delta outruns alignment.

The next part takes DELTA beyond matter in motion—into systems that learn, decide, and endure. What friction is to motion, stress is to behavior. What Bernoulli is to flow, coherence is to intelligence.

We now leave mechanics behind—not by abandoning physics, but by letting it finish its sentence.

Bernoulli does not tell us where energy goes. It tells us where alignment still holds.

Interlude — When Continuity Becomes Fragile

Up to this point, DELTA has spoken in the language of matter, motion, and flow. Pressure resolved. Boundaries aligned. Friction diffused. Bernoulli marked where coherence survived. Everything remained tangible—visible in bubbles, pipes, rivers, and wings.

But something subtle has already been exposed.

Coherence is not guaranteed.

Everywhere we looked, continuity survived only under specific conditions. Alignment had to propagate in time. Delta had to arrive gently enough to be shared. When those conditions failed, nature did not argue—it *jumped*. Films ruptured. Flow fragmented. Motion collapsed into heat and noise.

These jumps were not dramatic exceptions. They were the rule waiting behind continuity.

This interlude marks the moment where DELTA turns inward.

Because once we accept that continuity is conditional in fluids, we can no longer pretend it is guaranteed elsewhere. The same constraints that govern bubbles and friction must govern atoms, cells, minds, and meaning. The only difference is **scale and bandwidth.**

At very small scales, alignment bandwidth shrinks. Delta becomes starved. Resolution cannot spread smoothly. Systems no longer glide—they hop. What we call quantum jumps are not mysteries of nature; they are *consequences of coherence scarcity.*

At larger scales, some systems discover a workaround. They build internal loops. They recycle delta instead of exhausting it. They maintain resolution continuity against entropy. Those systems are called *living.*

And somewhere inside those living systems, alignment becomes aware of itself.

This is the turning point of the book.

From here onward, DELTA no longer explains *how things move.* It explains *why some things jump, some flow, and some stay alive long enough to ask questions.*

Randomness, life, and mind will no longer be treated as separate mysteries. They will appear as different outcomes of the same rule applied under different alignment constraints.

Small delta creates jumps. Sustained delta creates life. Reflected delta creates meaning.

With that, we step into next part — *From Quantum Jumps to Living Flow,* where continuity becomes precious, randomness becomes conditional, and the observer finally enters—not as a magician, but as a carefully organized participant in resolution.

The physics does not change. Only the stakes do.

Part IV — From Quantum Jumps To Living Flow

Overview

This part takes DELTA across its most fragile boundary: the point where continuity can no longer be assumed.

Until now, resolution unfolded smoothly when alignment and timing allowed it. In this part, we examine what happens when alignment bandwidth becomes scarce, internal pathways collapse, or delta can no longer spread continuously. The result is not mystery, but *discontinuity*—jumps, noise, and randomness that emerge as symptoms of failed coherence.

At the smallest scales, this appears as quantum behavior: not as fundamental unpredictability, but as resolution occurring without continuity. At intermediate scales, some systems escape this fate by building internal loops that recycle delta and preserve alignment. Those systems are living. At the most complex scales, alignment begins to reflect inward, producing awareness, meaning, and language—not as metaphysical additions, but as compressed records of resolved experience.

This part reframes randomness, life, and mind as outcomes of the same physical constraint applied under different conditions. Nothing new is introduced. DELTA remains the only driver. What changes is the system's capacity to sustain resolution over time.

Where continuity fails, jumps appear. Where continuity is maintained, life persists. Where continuity reflects, meaning emerges.

This part completes the transition from mechanics to existence—without leaving physics behind.

Chapter 11 — Small Delta, Big Jumps

When Continuity Runs Out

Quantum behavior has always been described as strange because it violates an expectation we rarely question: that reality should move smoothly.

We are comfortable with continuity. Objects travel along paths. Causes unfold into effects. Change spreads through space and time. When this expectation fails—when particles "jump," outcomes appear probabilistic, and trajectories vanish—we call it mystery.

But what if the mystery is not in nature, but in the expectation?

Everywhere else in this book, continuity has appeared as a conditional privilege. Fluids flow only when alignment bandwidth allows it. Boundaries hold only while ripple coherence survives. Motion persists only while delta can be resolved without spilling inward. When those conditions fail, nature does not hesitate. It jumps, fragments, diffuses.

Why should the smallest scales be any different?

At quantum scales, systems operate under extreme constraints. Delta is scarce. Spatial extent is minimal. Alignment pathways collapse almost immediately. There is simply no room for continuity to unfold. Resolution still must occur—but it cannot spread. It has nowhere to go.

So it jumps.

Quantum discontinuity is not an exotic feature added to reality. It is what resolution looks like when continuity is impossible. No path can form. No intermediate states can stabilize. The system moves directly between allowed configurations because that is the only resolution geometry available.

This reframing removes the need for mysticism without removing the strangeness. Quantum behavior remains unfamiliar, but it is no longer inexplicable. It follows the same rule already seen at larger scales—just under far harsher constraints.

Noise, uncertainty, and probability do not arise because nature is fundamentally random. They arise because alignment collapses faster than it can propagate. What we measure is not indeterminacy, but the visible trace of failed coherence.

The quantum world does not disobey physics. It obeys it too strictly.

Once this is understood, randomness stops being a philosophical threat and becomes a physical condition. Predictability returns where alignment bandwidth increases. Discontinuity fades as continuity becomes possible again. Nothing magical bridges the quantum and classical worlds—only capacity.

This chapter begins there: not by explaining quantum mechanics, but by explaining why smooth explanation itself runs out at small delta.

When delta cannot flow, resolution jumps.

And that simple fact will quietly dismantle much of what we have been taught to fear about the quantum world.

11.1 — Quantum Fluctuations as Delta-Starved Systems

At quantum scales, the defining limitation is not energy, not force, and not information. It is alignment bandwidth.

There is simply not enough spatial or temporal room for delta to spread.

In classical systems, when a small imbalance appears, alignment propagates outward. Neighboring regions adjust. Continuity forms.

Motion becomes smooth. Even turbulence, chaotic as it looks, is still a *continuous* redistribution of delta across many interacting pathways.

At quantum scales, those pathways do not exist.

A quantum system is delta-starved—not because delta is absent, but because it cannot be shared. Spatial extent is minimal. Temporal resolution is sharp. Any misalignment introduced immediately saturates the system's capacity to absorb it smoothly.

Resolution still must occur. But it cannot flow.

This is the key point: quantum fluctuations are not signs of excess energy or spontaneous chaos. They are minimum viable resolution events—the smallest adjustments a system can make when alignment cannot propagate.

In a delta-starved regime:

- There is no room to build gradients
- No neighboring structure to absorb misalignment
- No time window for gradual adjustment

The system is forced into discrete transitions because continuity would require bandwidth it does not have.

What we call a "quantum fluctuation" is not noise bubbling out of nothingness. It is the system repeatedly attempting to resolve unavoidable misalignment with almost no degrees of freedom. Each attempt is small, abrupt, and localized. There is no path between states because a path would require sustained alignment.

This also explains why quantum fluctuations persist even in apparent vacuum. Vacuum is not emptiness; it is a field with minimal structure. Delta exists—but the field cannot organize it continuously. Resolution appears as flicker, not flow.

Importantly, this behavior is not unique to quantum systems in principle. We see the same pattern whenever any system becomes delta-starved:

- Thin soap films rupture suddenly instead of deforming smoothly
- Brittle materials crack instead of bending
- Over-compressed systems snap instead of relaxing

Quantum systems simply live permanently at that edge.

From a DELTA perspective, quantum fluctuations are not exotic phenomena demanding new metaphysics. They are the natural outcome of resolution under extreme alignment scarcity. The system does the only thing it can do: jump between allowed configurations.

There is no hidden randomness here—only structural constraint.

Once this is understood, the next mystery becomes less threatening. If delta cannot spread and continuity cannot form, then classical notions like trajectories, paths, and smooth evolution must disappear.

That disappearance is not a failure of description. It is a necessary consequence.

And that is where we turn next: to discontinuity itself—not as an anomaly, but as resolution without continuity.

11.2 — Discontinuity Without Continuity

In classical physics, motion is inseparable from path. An object is expected to occupy every intermediate position between where it starts and where it ends. Change unfolds through continuity, and continuity gives us intuition.

At quantum scales, this expectation quietly collapses.

Not because nature chooses to be strange—but because continuity has no physical support.

When alignment bandwidth is sufficient, delta resolves by spreading. Intermediate states exist because the system can sustain them long enough to be real. But in delta-starved systems, any intermediate configuration would require time, space, and coupling that simply are not available. The system cannot *be* in between.

So it doesn't try.

Resolution occurs directly between allowed states. What we call a "jump" is not a leap across space; it is the absence of a sustainable intermediate alignment. There is no trajectory because a trajectory would be an act of continuous resolution—and continuous resolution is impossible under these constraints.

This is why quantum systems are described by states rather than paths.

A state is not a location in motion. It is a stable alignment configuration.

Between two such configurations, nothing meaningful can persist. Any attempt to interpolate collapses immediately into one state or the other. The system does not pass through forbidden or unstable alignments because those alignments cannot exist long enough to matter.

Discontinuity, then, is not randomness inserted into physics.

It is resolution occurring at the only points where alignment is viable.

This also explains why classical intuitions about velocity, acceleration, and causality break down. Those concepts presuppose continuity. They presuppose that change can be tracked moment by moment. In quantum regimes, change happens—but not in a trackable way. It is event-based, not trajectory-based.

Importantly, this does not mean that "anything can happen." The allowed states are tightly constrained. What disappears is not order, but the illusion of smooth passage between orders.

From the DELTA perspective, the rule is simple:

- Where alignment can propagate → continuity appears
- Where alignment cannot propagate → discontinuity

appears

Quantum systems live almost entirely in the second regime.

This reframing also dissolves the temptation to invoke observers as creators of reality. Measurement does not force a jump by mystical intervention. Measurement injects delta into a system that cannot distribute it continuously, leaving only one option: resolve into a new stable state.

The jump is not caused by observation. It is caused by insufficient alignment capacity.

Once this is seen, the drama drains away. Discontinuity is not a violation of physics. It is physics operating without the luxury of continuity.

The remaining confusion lies in what we call *noise* and *randomness*. If systems jump without paths, outcomes appear unpredictable. But unpredictability does not automatically mean fundamental randomness.

To see the difference, we must understand noise—not as chance, but as failed coherence attempting to resolve.

That is where we go next.

11.3 — Noise as Failed Coherence

Noise is usually treated as a primitive. It is assumed to be inherent, unavoidable, and irreducible—especially at small scales. When measurements fluctuate, when outcomes scatter, when signals refuse to stabilize, we shrug and say, *that's noise.*

Under DELTA, noise is not primitive. It is what unresolved alignment looks like.

Noise appears when a system is repeatedly asked to resolve delta but lacks the bandwidth to do so coherently. Each attempt leaves a partial trace. Those traces do not align into a smooth outcome or a stable state. They interfere, overlap, and fragment. What we observe is not randomness emerging from nothing, but coherence repeatedly failing to complete.

This distinction matters.

Randomness implies no structure. Noise implies structure that cannot settle.

In quantum systems, noise arises because alignment collapses before it can propagate. Each interaction—thermal, electromagnetic, measurement-induced—injects delta into a system that cannot distribute it smoothly. The system responds with abrupt, minimal resolution events. When these events accumulate without forming a stable alignment loop, the result appears as noisy behavior.

Importantly, noise is not evenly distributed. It has patterns, correlations, spectra. These are fingerprints of attempted alignment. Even quantum noise is not featureless. It carries information about coupling, boundary conditions, and timing. That information exists precisely because the system is *trying*—and failing—to maintain coherence.

Measurement amplifies this effect.

Measurement does not reveal noise that was already there, nor does it create reality out of nothing. Measurement introduces additional delta—interaction energy, coupling, constraint. In delta-starved systems, this extra demand overwhelms alignment capacity instantly. The system resolves by jumping, but the repeated injections prevent stabilization. Noise increases not because reality became uncertain, but because coherence was never allowed to settle.

This also explains why isolation reduces noise. Shielding, cooling, and decoupling do not remove randomness; they increase alignment bandwidth. With fewer disturbances and slower delta injection, the system has a chance—however brief—to sustain coherence. Noise drops because alignment starts to hold.

Seen this way, noise is diagnostic.

- High noise indicates repeated ATT violation
- Low noise indicates partial coherence survival
- Structured noise indicates constrained resolution pathways

Noise is what we hear when alignment tries to speak but is interrupted mid-sentence.

This reframing removes the temptation to treat noise as a metaphysical limit. It is not a wall imposed by nature. It is a symptom of insufficient coherence under demand.

And that leads directly to the most stubborn claim in modern physics: that randomness itself is fundamental.

To examine that claim honestly, we must separate *conditional unpredictability* from *ontological indeterminism*.

That is the task of the next section.

11.4 — Why Randomness Is Conditional, Not Fundamental

Randomness is often treated as the bedrock of modern physics. At small scales, we are told, outcomes are not just unpredictable in practice but unknowable in principle. Nature, it seems, rolls dice.

DELTA challenges this claim—not by denying unpredictability, but by relocating its cause.

Unpredictability does not require fundamental randomness. It requires insufficient coherence.

Whenever alignment bandwidth is too narrow to sustain continuous resolution, outcomes fragment into discrete possibilities. Which possibility is realized depends sensitively on micro-conditions that cannot be carried forward coherently. To an external observer, the result appears random. To the system, it is simply resolving under constraint.

Randomness, then, is not an ontological feature of reality.

It is a contextual outcome of alignment collapse.

This distinction becomes clear when coherence is ncreased. As coupling improves, disturbances are reduced, and timing is respected, predictability returns. Quantum systems cooled, isolated, or carefully driven begin to behave more regularly. Noise decreases. Correlations strengthen. The "random" begins to look structured.

If randomness were fundamental, no amount of coherence would matter. But it does.

What physics often calls randomness is actually conditional indeterminacy—indeterminacy that appears only when the system cannot sustain a continuous alignment narrative. The rules governing allowed states remain strict. What is lost is the ability to trace a smooth cause–effect path between them.

This is why probability enters the picture.

Probability is not a property of nature deciding outcomes. It is a tool for observers dealing with systems whose resolution capacity has collapsed below continuity. When alignment cannot carry history forward, outcomes must be described statistically. Probability is not explanation; it is post-event compression.

Seen this way, determinism and randomness are not opposites. They are endpoints of a spectrum governed by alignment bandwidth.

- High coherence → continuity → determinism
- Low coherence → discontinuity → apparent randomness

Neither regime is more "real" than the other. They are consequences of the same physical rule applied under different constraints.

This reframing also dissolves the philosophical anxiety around free will, causality, and predictability. Nature is not undecided. It is overconstrained. When constraints prevent smooth resolution, the system jumps. When they allow it, the system flows.

Randomness is what we see when we ask continuity to exist where it physically cannot.

Understanding this prepares the ground for the next step. If randomness is conditional and probabilities are retrospective, then the mathematical objects we use to describe quantum systems—wavefunctions, amplitudes, distributions—must also be reinterpreted.

They are not pictures of motion. They are maps of allowed resolution.

That is where we turn next.

11.5 — Probability as Post-Jump Accounting

Probability enters physics not at the moment of action, but after continuity has failed.

This is the crucial inversion.

We are taught to believe that probability governs what will happen. Under DELTA, probability records what could have happened once smooth resolution was no longer possible. It is not a driver of events. It is a bookkeeping language adopted when trajectories disappear.

When a system can sustain alignment, it does not need probability. Cause flows into effect. History propagates. Outcomes are traceable. The moment alignment collapses—when delta cannot spread and resolution must jump—the system loses the ability to carry a continuous narrative forward.

At that point, probability appears.

Probability is not uncertainty in nature. It is uncertainty in continuity.

A quantum jump does not select an outcome *because* probabilities demand it. The jump occurs because resolution must occur somewhere among allowed states. After the jump, we assign probabilities to describe the distribution of outcomes across many similar collapses. The probabilities do not cause the events; the events force us to use probabilities.

This is why probabilities in quantum mechanics are always statistical and ensemble-based. No single jump "consults" a probability. Only after many jumps do stable distributions emerge. Probability is post-resolution compression—a way to summarize repeated discontinuities when we cannot track individual causal chains.

Seen this way, the wavefunction stops being mysterious.

It does not represent a particle smeared across space. It represents an allowance landscape—a map of where resolution is permitted if and when continuity fails.

Before a jump, the wavefunction encodes constraints, couplings, boundary conditions, and alignment limits. After a jump, it collapses because the bookkeeping problem has changed. One allowed configuration has been realized; the rest are no longer relevant for that event.

Nothing physical collapses. Only the description updates.

This also explains why probabilities change when experimental arrangements change. Altering boundaries, coupling, or measurement conditions reshapes the allowance landscape. Different deltas are injected. Different alignments are stressed. The distribution of jumps shifts accordingly—not because reality is fickle, but because resolution conditions have changed.

Probability, then, is not a fundamental ingredient of nature. It is a language we adopt when continuity is unavailable.

Whenever alignment bandwidth increases—through isolation, coherence, or scale—probability recedes. Trajectories reappear. Deterministic descriptions return. The same system that demanded probability yesterday may permit continuity tomorrow under gentler conditions.

This is not a flaw in physics. It is physics being honest about its limits.

Once probability is seen as post-jump accounting rather than pre-event law, the divide between quantum and classical behavior softens. There is no ontological gap—only a bandwidth transition.

And that transition, from jump-dominated behavior back to flow, is precisely what we examine next.

11.6 — From Quantum Jumps to Classical Flow

The transition from quantum jumps to classical flow is often presented as a mystery—something that happens vaguely "when systems get large," or when observers stop paying attention. Under DELTA, the transition is neither vague nor mysterious. It is structural.

Continuity reappears when alignment bandwidth increases.

At small scales, delta cannot spread. Resolution collapses into jumps because there is no room—spatially, temporally, or relationally—for alignment to propagate. As scale increases, something critical changes: degrees of freedom multiply. The system gains neighbors, coupling channels, buffering time, and internal pathways through which delta can distribute.

This does not eliminate delta. It changes how delta resolves.

As alignment bandwidth grows, resolution no longer needs to occur all at once. Intermediate configurations can stabilize briefly. Micro-adjustments can accumulate instead of collapsing. What was previously a jump becomes a sequence. What was previously noise becomes flow.

This is why classical behavior emerges gradually, not abruptly.

There is no sharp boundary where quantum rules switch off and classical rules switch on. There is a crossover regime, where partial coherence exists. In this regime, systems exhibit mixed behavior—mostly continuous, occasionally jumping. Decoherence is not a magical process that "destroys quantum effects." It is the redistribution of alignment into many interacting channels, making jumps unnecessary.

When delta is shared widely enough, the system no longer needs to resolve discontinuously.

Importantly, nothing fundamentally new appears at the classical level. No new forces are added. No hidden variables awaken. What appears is capacity—the capacity to sustain alignment over time.

This reframes classical determinism itself. Determinism is not a deep metaphysical truth about nature. It is a luxury afforded by bandwidth. When alignment can propagate reliably, outcomes become traceable. When it cannot, they fragment.

From this perspective, the classical world is not more "real" than the quantum world. It is simply more forgiving. It tolerates gradients. It absorbs disturbance. It smooths delta across space and time.

The same delta operates in both regimes. The same rule governs resolution. Only the geometry of alignment changes.

This insight dissolves the artificial divide between micro and macro physics. Quantum mechanics is not a different universe. It is what physics looks like when coherence is scarce. Classical mechanics is what physics looks like when coherence is abundant.

And once continuity becomes sustainable, something new becomes possible.

Resolution can loop.

It can circulate, buffer itself, and persist against decay. It can build internal pathways that maintain alignment even in the presence of disturbance. That capacity marks a profound shift—not just from quantum to classical, but from non-living to living.

That is where this journey naturally leads next.

From Jumps to Continuity That Remembers

Quantum jumps are not failures of explanation. They are what resolution looks like when it cannot remember itself.

In a delta-starved system, alignment resolves and vanishes instantly. There is no memory, no buffering, no persistence. Each jump stands alone. Probability steps in afterward to summarize what could not be carried forward. Randomness appears—not as a fundamental truth, but as a side effect of amnesia.

What this chapter has shown is simple but unsettling: continuity is not guaranteed by nature. It must be *earned.*

When alignment bandwidth increases, jumps soften into sequences. Noise condenses into patterns. Probability retreats as coherence returns. Physics does not change its rules—it merely gains the ability to keep a story going.

But even classical flow, for all its smoothness, still forgets. It dissipates. It relaxes. It runs downhill.

Something else is required for continuity to do more than pass through.

The next step is not just sustained resolution, but sustained resolution with internal loops—systems that trap delta, recycle alignment, and resist decay long enough to matter.

That step is called life.

In the next chapter, we cross the most important threshold in this book: from resolution that happens, to resolution that stays.

When delta is too small to spread, resolution jumps instead of flowing.

Chapter 12 — Life as Sustained Delta Resolution

When Resolution Refuses to End

Life has always been treated as an exception.

Physics explains motion, chemistry explains reactions, biology explains organisms—and somewhere between these disciplines, life is quietly handed over to mystery. We invent checklists, argue about definitions, and search for special ingredients, as if life were a substance waiting to be discovered.

But nothing special ever appears.

Atoms in a living body are not unique. Energy in a living system is not exotic. The laws governing a cell are the same laws governing a stone.

So where does life actually begin?

DELTA offers a blunt answer: life begins where resolution does not terminate.

In non-living systems, delta resolves and disappears. A pressure difference equalizes. A temperature gradient relaxes. A disturbance spreads, dissipates, and the story ends. The system reaches equilibrium—or fractures—and moves on.

Living systems do something quietly radical. They refuse to finish resolving.

They take incoming delta and loop it back into themselves. They prevent alignment from running downhill to rest. They maintain internal misalignment just enough to stay active, organized, and responsive. Life is not equilibrium plus energy. It is equilibrium continuously postponed.

This is not metaphor. It is mechanics.

A flame also resists equilibrium, but only briefly. A hurricane maintains structure, but only while external conditions cooperate. Living systems persist across changing environments because they build internal pathways that recycle resolution rather than exhaust it.

This is the decisive shift.

Life is not motion. Life is not reaction. Life is sustained resolution with memory.

Memory here is not mental. It is structural—the ability of a system to carry alignment forward through time without collapsing or dissipating it entirely. This is why life feels directional. Why it grows, repairs, adapts, and resists decay. It is not striving. It is organized delay.

Seen this way, life stops being a philosophical riddle and becomes a physical threshold. Cross that threshold, and matter begins to behave differently—not by breaking laws, but by using them continuously.

This chapter begins there.

Not by asking what life is made of, but by asking why, in some systems, resolution simply refuses to end.

12.1 — What Separates Living from Non-Living

The boundary between living and non-living is often searched for in *things*: DNA, cells, metabolism, reproduction, consciousness. Each candidate captures something important—and each one fails as a universal divider.

Viruses have genes but are not alive on their own. Crystals grow but do not live. Fire consumes energy and propagates but quickly dies.

What all these attempts miss is not a component, but a behavioral regime.

From a DELTA perspective, the separation is clean:

Non-living systems resolve delta once. Living systems resolve delta repeatedly without closure.

A non-living system experiences imbalance, resolves it, and settles. The resolution may be fast or slow, violent or gentle, but it is terminal. Once alignment is restored—or destroyed—the process ends. There is no insistence on continuation.

A living system, by contrast, is defined by non-terminal resolution.

It does not seek equilibrium. It actively avoids it. Not by violating physics, but by reinjecting delta into itself before resolution completes. Every moment of apparent balance is provisional. Every resolved gradient is immediately replaced by another, internally generated one.

This is why energy alone cannot define life.

Energy can move, react, dissipate, and reorganize matter indefinitely. But without internal loops, energy always runs downhill. Living systems do not stop that descent—they delay it locally by structuring how delta enters, circulates, and exits.

This also explains why motion is insufficient. A rock rolling downhill moves. A river flows. Neither lives. Their motion resolves delta in one direction only. They do not preserve the conditions that allow motion to continue once the gradient is gone.

Living systems do preserve those conditions.

They maintain temperature gradients, chemical gradients, electrical gradients—not accidentally, but structurally. They do not wait for delta to arrive. They manufacture it internally through organized processes.

The difference, then, is not complexity, intelligence, or purpose. It is commitment to continuation.

Non-living systems do not care whether resolution ends. Living systems are built so that it cannot.

This reframing also dissolves borderline cases. Anything that can sustain internal delta loops, repair those loops, and adapt their pacing qualifies as living—at least partially. Anything that cannot will eventually relax into non-life, regardless of how complex it looks.

Life is not a label we assign. It is a regime of dynamism.

Understanding this sets the stage for the core mechanism behind life's endurance: internal delta loops—the structures that make non-terminal resolution physically possible.

That is where we turn next.

12.2 — Internal Delta Loops

If life is sustained resolution, then internal delta loops are the machinery that makes it possible.

In non-living systems, delta enters from outside, resolves once, and disappears. A stone warms in sunlight and cools at night. A gas expands, equilibrates, and stops. Even complex non-living systems depend on external gradients to remain active. When those gradients vanish, so does the behavior.

Living systems break this pattern by doing something subtle but decisive: they close the loop.

An internal delta loop is not a metaphor for feedback or regulation. It is a physical circulation of misalignment. Delta enters the system, is partially resolved, and instead of being exhausted, is redirected to create new internal gradients. These gradients then drive further processes, which in turn regenerate delta elsewhere in the system.

Resolution feeds the next imbalance.

This looping is what turns a one-time event into a process.

Consider metabolism at its most basic level. Nutrients enter a cell not merely to release energy, but to restructure internal alignment—maintaining chemical, electrical, and mechanical gradients. Those gradients power transport, synthesis, repair, and signaling. The waste products carry away misalignment that can no longer be integrated, preserving internal coherence.

The loop is asymmetric and directional, but it is not runaway. It is paced. ATT is respected internally. Too fast, and the system overheats, destabilizes, or poisons itself. Too slow, and gradients decay. Life survives in the narrow corridor where internal delta circulation matches the system's capacity to reorganize.

This is why feedback in living systems is physical, not informational.

Signals are not instructions. They are localized deltas—small, targeted misalignments injected into specific pathways. Hormones, neurotransmitters, and electrical impulses do not tell the system what to do. They perturb it in precise places, allowing resolution to propagate along established loops.

Internal delta loops also explain why life is resilient.

When one pathway fails, others compensate. When external conditions change, internal circulation adapts. The system does not depend on a single gradient, but on a network of mutually sustaining imbalances. As long as enough loops remain intact, resolution continues.

Break the loops, and life collapses quickly.

This is why poisons are effective not because they remove energy, but because they interrupt circulation. This is why starvation kills not by immediate energy loss, but by gradual collapse of internal gradients. And this is why repair is central to life: maintaining loops is more important than creating new ones.

Life persists not because it is strong, but because it is architected for circulation.

Internal delta loops transform physics from a one-way descent into a managed cycle. They do not defeat entropy; they negotiate with it continuously.

With this mechanism in place, we can now re-express a familiar biological concept in physical terms: metabolism—not as fuel consumption, but as alignment maintenance.

That is the next step.

12.3 — Why Metabolism Is Physical Alignment

Metabolism is usually described as chemistry with purpose: reactions that release energy, synthesize molecules, and keep the organism alive. This description is accurate—but incomplete. It hides the deeper role metabolism plays in the physics of life.

Metabolism is not primarily about energy extraction. It is about alignment maintenance.

From a DELTA perspective, every metabolic reaction serves one of two functions: either it builds alignment where coherence is needed, or it exports misalignment where coherence would otherwise collapse. Energy is involved in both, but energy is not the objective. Alignment is.

Nutrients enter a living system carrying structured delta. They are not just sources of calories; they are alignment carriers—molecules arranged in ways that can be reorganized into internal gradients. Glucose, oxygen, amino acids, ions—each participates in sustaining chemical, electrical, and mechanical misalignments that drive circulation within the organism.

When these materials are metabolized, their internal alignment is not simply destroyed. It is redistributed. Some of it reinforces existing gradients. Some of it is converted into motion, synthesis, or repair. The remainder—misalignment that cannot be integrated—is expelled as waste.

Waste is not failure. It is alignment that no longer fits.

This reframing clarifies why metabolism must be continuous. Gradients decay naturally. Membranes leak. Structures fatigue. Without constant alignment repair, internal loops would collapse. Metabolism keeps resolution incomplete on purpose—always replacing lost coherence before it vanishes entirely.

It also explains why metabolic rate matters. Too rapid, and alignment is injected faster than internal ATT allows. Heat accumulates. Toxic intermediates build up. The system destabilizes. Too slow, and gradients fade. The system drifts toward equilibrium. Life exists only in the narrow window where alignment input matches alignment capacity.

This is why metabolic disorders are alignment disorders. Diabetes, hypoxia, mitochondrial dysfunction—these are not energy shortages in the crude sense. They are failures to maintain or distribute alignment coherently. Energy may be present, but it is misrouted, mistimed, or unusable.

Metabolism, then, is not a fire burning inside the body. It is a carefully paced alignment choreography.

Seen this way, familiar biological distinctions sharpen. Aerobic metabolism sustains higher alignment bandwidth by coupling to efficient external sinks. Anaerobic metabolism survives by accepting lower coherence and higher misalignment export. Growth increases alignment capacity. Aging reduces it.

Life does not consume energy to live. It spends energy to stay aligned.

With metabolism understood as physical alignment maintenance, the next question becomes unavoidable: how does time behave inside such systems? Why does living time feel directional, rhythmic, and personal—unlike the neutral time of mechanics?

That question leads us directly to the role of time inside living systems.

12.4 — Time Inside Living Systems

Time inside a living system does not behave like time in a mechanical one.

A clock ticks whether anything happens or not. A pendulum swings until friction stops it. Their time is external, imposed, indifferent to internal state. Living systems experience time differently—not subjectively first, but physically.

This difference arises from how alignment is sustained.

In non-living systems, time is simply the duration over which delta resolves. Once resolution ends, time becomes irrelevant to the system's behavior. In living systems, resolution never fully completes. Internal delta loops keep alignment active, and time becomes embedded in the circulation itself.

Living time is not measured by seconds. It is measured by cycles.

Heartbeat, respiration, circadian rhythms, metabolic turnover, cellular repair—each is a pacing mechanism that regulates how quickly delta is allowed to resolve. These rhythms are not add-ons. They are ATT regulators—internal clocks that match resolution rate to alignment capacity.

This is why time inside living systems feels directional and structured. Processes must occur in the right order, at the right speed, with the right spacing. Too fast, and coherence collapses. Too slow, and gradients decay. Living time is therefore negotiated, not absolute.

ATT becomes system-specific and dynamic.

A child's system tolerates rapid alignment turnover. Growth is fast, repair is quick, rhythms are flexible. With age, alignment bandwidth narrows. Repair slows. Rhythms stiffen. The same external minute contains less usable resolution. Aging is not time passing faster; it is alignment capacity shrinking.

This also explains why stress alters time perception and physiology simultaneously. Stress injects large delta into internal loops. ATT is exceeded. Processes desynchronize. Subjectively, time feels distorted; physically, coherence erodes. The sensation and the mechanism are the same phenomenon viewed from different angles.

Even sleep fits cleanly into this picture. Sleep is not inactivity. It is alignment rebalancing—a period where external delta input is minimized so internal loops can repair phase errors, restore gradients, and resynchronize timing. Deprive a system of this recalibration, and time inside it fragments.

Living systems do not live *in* time. They manufacture time through paced resolution.

This is why life cannot be sped up arbitrarily without cost. Faster is not better. Faster means higher delta per unit time, and beyond a point, alignment cannot keep up. Damage accumulates. Coherence leaks. The system ages prematurely.

Time, for life, is not a neutral dimension. It is a resource shaped by alignment capacity.

With this understanding, we can now clarify a common confusion: if living systems avoid equilibrium, how do they remain stable at all? Why don't internal loops spiral into chaos?

The answer lies in a special kind of balance—stability without stillness.

That is where we turn next.

12.5 — Stability Without Stillness

Living systems are often described as being "in balance," but this phrase is dangerously misleading. Balance suggests rest. Stillness. Completion. Life has none of these.

A living system is stable precisely because it never becomes still.

In non-living systems, stability means equilibrium. Forces cancel. Gradients flatten. Motion ceases. The system becomes predictable—and inert. Any disturbance is resolved by returning to rest.

Living systems operate under a different logic. They remain stable by never allowing full resolution. Gradients are maintained just enough to keep processes running. Misalignment is tolerated, redirected, and used. Stability is achieved not by eliminating delta, but by keeping delta circulating within safe bounds.

This is why life feels paradoxical to force-based thinking. From the outside, a living organism looks calm, ordered, even static. Inside, it is a storm of reactions, transports, adjustments, and corrections. What appears as calm is actually dynamic coherence—many processes moving, but moving in phase.

Health, in this sense, is not the absence of disturbance. It is the margin of coherence.

A healthy system can absorb fluctuations without losing alignment. It can tolerate spikes in delta—temperature changes, emotional stress, physical exertion—because its internal loops have slack. ATT is respected even under load. Resolution speeds up temporarily without tearing the system apart.

Illness begins when that margin narrows.

Chronic stress, toxins, malnutrition, aging—all reduce alignment bandwidth. Internal loops become brittle. Small disturbances now exceed ATT. Processes fall out of sync. What once would have been absorbed becomes damaging. Symptoms appear not because something new has gone wrong, but because the system has lost its buffer.

This reframing also clarifies why aggressive intervention often backfires. Forcing a system toward "normal" values—pressure, glucose, mood, performance—without restoring alignment capacity simply increases delta faster than the system can handle. Temporary correction is followed by deeper instability.

Living stability is therefore not control. It is adaptive tolerance.

The system must be allowed to move, fluctuate, and even wobble—so long as coherence is preserved. This is why rigidity is a sign of fragility, not strength. Flexible systems survive. Rigid ones snap.

Life maintains form by flowing through it, not by freezing into it.

Understanding this prepares us for the final boundary of life. If stability depends on circulating alignment, then death is not mysterious either. It is not the loss of energy, nor the disappearance of matter.

It is the point where circulation fails.

That is where we go next.

12.6 — Death as Loss of Resolution Continuity

Death is often described as disappearance: of energy, of function, of identity. None of these descriptions withstand physical scrutiny. Energy remains. Matter remains. Even many chemical processes continue briefly after life ends.

What vanishes is continuity.

From a DELTA perspective, death is not an event—it is a threshold crossing. It is the moment when internal delta loops can no longer sustain resolution. Alignment still exists locally, but it can no longer circulate. The system loses its ability to keep misalignment moving without collapsing into rest.

In life, resolution is postponed. In death, postponement fails.

This failure is not sudden in all systems. It can be gradual or abrupt, localized or systemic. But the mechanism is the same: alignment bandwidth drops below the minimum required to keep loops closed. Once that happens, resolution completes instead of circulating.

This is why death is irreversible.

You can add energy to a dead system. You can add heat, electricity, nutrients. What you cannot restore is the architecture that routes delta back into itself.

The loops are gone.

Membranes lose selective coherence. Gradients flatten irreversibly. Timing desynchronizes beyond repair. ATT is violated everywhere at once—not because delta is large, but because capacity is gone. Resolution finishes its work and does not restart.

Importantly, nothing dramatic is required. No explosion, no depletion, no mystical departure. Death can occur quietly when internal loops decay just enough that circulation cannot resume. The system slips into equilibrium—not as a choice, but as a consequence.

This also explains why revival is only possible near the boundary.

As long as some loops remain intact—some gradients maintained, some timing preserved—resolution continuity can be restarted. Cooling, defibrillation, oxygenation work not because they inject life, but because they slow resolution long enough for alignment to re-form. Cross the threshold fully, and no intervention can reverse it.

Death, then, is not the opposite of life. It is life's completion.

The same physics governs both. Life delays resolution. Death allows it to finish.

Seen this way, death loses its metaphysical weight without losing its seriousness. It is not a punishment, a failure, or a mystery. It is a physical boundary—clear, final, and universal.

Understanding this boundary allows us to make one last clarification before moving on. If life is sustained resolution, it may feel extraordinary—but it is not special in the cosmic sense.

It is simply rare.

That recognition closes this chapter.

12.7 — Life Is Not Special—It Is Rare

Once life is understood as sustained delta resolution, its place in the universe becomes both clearer and humbler. Life is no longer a miracle that defies physics, nor a privileged outcome written into the fabric of reality. It is a narrow dynamical regime—possible, but difficult to maintain.

The universe does not resist life. It simply does not optimize for it.

Everywhere, delta resolves. Gradients flatten. Systems relax. Equilibrium is the default trajectory. Life survives only where conditions allow internal loops to form and persist—where alignment can be circulated without exhausting itself too quickly.

That requirement is severe.

To sustain internal delta loops, a system must satisfy multiple constraints simultaneously: access to structured inputs, efficient waste export, protective boundaries, timing mechanisms, and sufficient alignment bandwidth. Remove any one of these, and resolution completes. The system drifts back toward non-life.

This is why life clusters. It appears where conditions are just right—not because the universe is tuned for biology, but because biology is tuned to exploit rare coherence windows. Water-rich environments, moderate temperatures, stable energy flows, and protective atmospheres are not coincidences. They are coherence enablers.

Life is not inevitable. It is opportunistic.

This reframing also dissolves anthropocentric temptation. Life does not sit at the center of physics. It sits at the edge—balanced precariously between runaway dissipation and rigid equilibrium. Its persistence is impressive not because it violates entropy, but because it negotiates with it continuously.

This perspective changes how we think about fragility. Life's vulnerability is not a flaw. It is a consequence of its operating regime. Systems that rely on sustained alignment must remain adaptable, repairable, and responsive. When they lose those traits, they do not degrade gracefully—they collapse.

Yet within this fragility lies power. Life can sense, adapt, and reorganize precisely because it maintains internal delta. It is sensitive by necessity. That sensitivity, extended inward, becomes awareness.

This brings the argument to its final turn.

If life is resolution that stays, then mind is resolution that knows it is staying.

The next chapter crosses that threshold—not into metaphysics, but into the physics of awareness, meaning, and the observer as an organized participant in dynamism.

That is where DELTA finally meets the mind.

From Sustained Flow to Reflective Flow

Life completes the first great escape from discontinuity.

Where non-living systems resolve and finish, living systems loop. Where quantum systems jump without memory, living systems remember through structure. Resolution does not end; it circulates, repairs itself, and postpones equilibrium long enough to matter.

But something crucial is still missing.

A living system can sustain alignment without knowing that it does so. Cells metabolize flawlessly without understanding metabolism. Organs coordinate without awareness. Life persists—but it does not yet reflect.

The next threshold is not about more energy, more complexity, or better chemistry. It is about internal coherence becoming visible to itself.

When alignment is not only sustained but monitored, when resolution is not only looped but referenced, when internal delta is not only managed but felt,

something new appears.

Not a soul. Not a ghost. Not an observer standing outside reality.

But an organized resolution system that can sense its own alignment state.

Mind does not arrive as an add-on to life. It emerges when life's internal loops become sufficiently coherent to reference themselves.

Meaning does not precede alignment. It is compressed *after* alignment stabilizes.

Language does not reveal truth. It interfaces between resolved experiences.

The next chapter does not abandon physics. It tightens it further.

From here on, DELTA explains not just how things persist, but how persistence becomes awareness—and how awareness turns resolution into experience.

That is where we go next.

Life is not energy in motion. It is alignment that refuses to finish resolving.

Chapter 13 — Mind, Awareness, and Meaning

When Resolution Begins to Notice Itself

Mind has always been treated as an intruder in physics.

Matter moves, energy flows, fields interact—and suddenly, somewhere in this orderly picture, awareness appears. We respond by pushing it outside the system: into philosophy, psychology, spirituality, or mystery. The observer becomes a ghost hovering over reality, watching but not belonging.

DELTA does not allow that escape.

If all dynamism is driven by delta resolving under alignment constraints, then mind cannot be an exception. It must be a configuration of the same process, not a departure from it.

The difficulty lies in a quiet shift we rarely articulate: at some point, resolution stops being blind.

Up to this point in the book, resolution flowed, looped, and persisted. Life delayed equilibrium by circulating alignment internally. But even living systems can operate without awareness. Bacteria metabolize flawlessly. Plants track light and water without noticing themselves.

Mind appears only when internal alignment becomes self-referential.

This is not metaphor. It is structure.

A system becomes a mind when its internal resolution loops are coherent enough to monitor their own alignment state. When resolution does not just happen, but is *registered* within the system as a condition. Awareness is not a thing added to life; it is what happens when life's internal loops cross a coherence threshold.

This reframing dissolves several long-standing confusions at once.

The observer does not stand outside reality. The observer is a resolution system operating with memory.

Observation is not interference. It is participation with consequences.

Awareness is not intelligence. It is coherence visibility.

Once this is seen, meaning, memory, language, and choice stop appearing magical. They become downstream effects—tools developed by a system that can track its own state while remaining inside the flow of resolution.

The mind is not a mirror of the world. It is a compressed continuation of it.

This chapter begins there: not by asking what mind is made of, but by asking what changes when resolution starts noticing itself.

That single shift—small in wording, enormous in consequence—will carry us through observerhood, meaning, and language without leaving physics behind.

From here on, experience itself becomes explainable.

13.1 — The Observer as an Organized Resolution System

The observer has long been treated as a problem.

In classical physics, the observer is ignored. In quantum mechanics, the observer becomes suspiciously powerful—collapsing wavefunctions, deciding outcomes, bending reality by looking at it. Both treatments share the same mistake: they place the observer outside the system being described.

DELTA removes that possibility entirely.

An observer is not a spectator. An observer is an organized resolution system embedded in the same dynamism it observes.

This means observation is not a passive act. It is a physical interaction—delta meeting delta—within a structured, self-sustaining system. When an observer "observes," alignment is not created or destroyed. It is redirected through internal loops that register change.

This reframing dissolves the illusion of observer-created reality. Reality does not wait to be noticed. Systems resolve delta continuously whether anyone is watching or not. What observation changes is not the world, but how resolution propagates through the observer.

An observer is defined by three features:

1. Internal coherence sufficient to sustain resolution loops
2. Memory structures that carry alignment forward
3. Reference pathways that allow the system to compare current alignment with prior states

Together, these allow the system to experience.

Experience is not raw sensation. It is resolution registered against itself.

This also clarifies why observers disturb what they observe. Measurement is not magical; it is intrusive. Introducing an observing system necessarily injects delta—coupling, energy exchange, constraint. In systems with limited alignment bandwidth, this injection forces resolution to jump. The observer did not create the outcome; it participated in the resolution.

At larger scales, this participation is negligible. At smaller scales, it dominates. The difference is not philosophical. It is bandwidth.

Importantly, the observer does not control outcomes. It does not choose reality. It navigates within constraints imposed by its own alignment capacity and the system it interacts with. Observation is an act of being affected.

This reframing also explains why observation can feel effortless or exhausting. When internal coherence is high, alignment flows smoothly. Observation feels clear, calm, and immediate. When coherence is low—due to overload, stress, or fragmentation—observation becomes noisy, distorted, or impossible.

The observer is not divine. It is fragile.

Understanding the observer as an organized resolution system removes the last refuge of dualism. Mind and world are not separate. They are coupled flows of delta, intersecting through structured interaction.

With this foundation in place, we can now ask a sharper question: what exactly allows an observer to *notice* rather than merely react?

That question leads directly to awareness itself—not as intelligence or consciousness, but as internal coherence made visible.

13.2 — Awareness as Internal Coherence

Awareness is often confused with intelligence, attention, or consciousness. It is treated as something a system *has*, rather than something a system *maintains*. Under DELTA, awareness is neither a substance nor a capability.

Awareness is a state of internal coherence.

An organized resolution system becomes aware when its internal loops are sufficiently aligned that changes in alignment are detectable within the system itself. Nothing mystical occurs. No new entity appears. The system simply reaches a point where resolution leaves a trace that can be compared against prior resolution.

Awareness is not thinking. It is registration.

A system can be intelligent—capable of complex processing, pattern recognition, and response—without being aware. Machines routinely demonstrate this. Reflexes in living organisms operate without awareness. Awareness arises only when internal coherence allows the system to hold a stable reference while resolution continues.

This is why awareness collapses under overload.

When delta exceeds internal ATT, alignment fragments. Internal loops fall out of sync. The system can still react, but it cannot *notice* its own state. Awareness fades not because consciousness leaves, but because coherence has dropped below the visibility threshold.

This also explains the narrowness of awareness. Awareness is not global. It is localized to regions of high internal coherence. Attention is not effortful focus; it is alignment allocation. When coherence is concentrated along certain loops, those loops become visible. Others fade into background activity.

Meditative calm, flow states, and deep focus all share the same physical signature: reduced delta injection, synchronized internal timing, and widened coherence margins. Awareness expands not because the mind opens, but because alignment stabilizes.

Conversely, anxiety, distraction, and fatigue fragment awareness. Too many deltas compete. Timing desynchronizes. The system loses its ability to register itself coherently.

This reframing removes the need for metaphysical consciousness. Awareness does not float above processes. It emerges within them when conditions allow.

Awareness is also dynamic. It waxes and wanes with physiological state, environment, and demand. Sleep, anesthesia, and injury reduce alignment bandwidth. Awareness fades. Nothing mysterious departs. The system simply cannot sustain self-registration.

Awareness, then, is not the driver of experience. It is the window through which experience becomes visible.

With awareness understood as internal coherence, the next step follows naturally. Once alignment is registered, it can be compressed. Patterns can be abstracted. Meaning can be formed.

That compression—what we call meaning—is not present in raw experience. It appears *after* alignment stabilizes.

That is where we go next.

13.3 — Meaning as Post-Alignment Compression

Meaning is often mistaken for something embedded in the world—signals waiting to be decoded, truths hidden inside events. Under DELTA, meaning is neither discovered nor received.

Meaning is constructed after alignment stabilizes.

Experience arrives first as raw resolution: changes in internal alignment triggered by interaction with the world. Before coherence settles, there is sensation but no meaning. The system is affected, but it has not yet *understood* anything. Understanding begins only when internal coherence holds long enough for comparison.

Meaning is compression.

Once alignment is stabilized across internal loops, the system condenses that distributed resolution into a manageable form—a pattern, a label, a narrative. This compression reduces detail, discards noise, and preserves only what is useful for future navigation. Meaning is therefore always lossy, contextual, and provisional.

This explains why meaning never precedes experience.

You cannot assign meaning to something you have not aligned with. When people try, they project expectation instead of understanding. Genuine insight occurs when alignment snaps into coherence, and compression follows automatically. The "aha" moment is not discovery—it is coherence crystallizing.

Meaning also feels powerful because it saves bandwidth. Instead of re-resolving the full complexity of an experience each time, the system stores a compressed representation. This allows faster response, planning, and communication. Meaning is not truth; it is efficiency.

This is why meaning changes with context.

As alignment conditions shift—emotionally, physiologically, socially—the same experience compresses differently. What once felt meaningful may later feel empty. What once felt trivial may suddenly carry weight. The world did not change. Alignment did.

This reframing dissolves the anxiety around subjective meaning. Meaning does not need to be universal to be valid. It needs only to be fit for the alignment state that produced it.

It also clarifies why meaning can mislead.

Because compression discards detail, it can oversimplify. Stored meanings can be applied outside their original context, producing distortion. Trauma, ideology, and dogma are not failures of intelligence; they are compressed meanings applied without re-alignment.

Meaning, then, is not something to worship. It is a tool produced by coherence.

Understanding this prepares the ground for language. Language is not where meaning lives. It is where meaning is exchanged—an interface built to transfer compressed alignment between systems.

That interface, and its limitations, are the subject of the next section.

13.4 — Memory and Continuity of Self

Memory is often treated as storage—a warehouse of past experiences stacked somewhere in the brain. Under DELTA, this picture is misleading. Memory is not where experiences are kept. Memory is how alignment persists.

Memory is coherence with duration.

When a system resolves delta and then fully relaxes, nothing remains. The event is over. In living, aware systems, resolution leaves behind structural traces—changes in pathways, sensitivities, thresholds, and timing. These traces bias future resolution. They make some alignments easier to re-enter and others harder to reach.

That bias *is* memory.

Memory does not record the past as it was. It records how the system was shaped by resolving it. This is why memory is reconstructive, not archival. Each recall is a fresh alignment event guided by existing coherence patterns. What feels like retrieval is actually re-resolution along familiar paths.

This also explains the continuity of self.

The "self" is not a thing carried forward unchanged. It is a stabilized resolution pattern—a recurring configuration of alignment that persists because internal loops repeatedly reinforce it. As long as these loops remain coherent, the system feels like the same self from moment to moment.

Identity is not fixed. It is maintained.

When alignment patterns shift gradually, identity feels continuous. When they shift abruptly—through trauma, injury, or extreme stress—continuity fractures. The system is forced into new resolution pathways faster than coherence can adapt. Memory becomes unreliable. The sense of self destabilizes. This is not psychological weakness; it is physical misalignment.

Trauma, in this framework, is not stored pain. It is coherence disruption.

A traumatic event injects delta beyond internal ATT. Resolution fragments instead of stabilizing. Compression occurs prematurely or incompletely. The resulting memory is not integrated; it remains a source of recurrent misalignment. When similar deltas appear later, the system is pulled back into partial resolution loops, reliving rather than remembering.

This also clarifies why healing requires more than explanation. Insight alone cannot repair alignment. The system must be allowed to re-resolve safely, rebuilding coherence gradually. Time, rhythm, safety, and repeated gentle exposure matter because they respect ATT.

Memory, then, is not a record of who you were. It is the mechanism by which you remain someone at all.

Understanding memory as alignment persistence dissolves the illusion of a permanent self without denying lived continuity. The self exists as long as coherence holds—and changes as coherence reshapes itself.

With memory and self clarified, one final layer remains: how these compressed alignments are exchanged between systems. That exchange is not experience itself, but language—the interface that allows one organized resolution system to perturb another.

That interface, and its limits, are next.

13.5 — Language as Interface, Not Truth

Language feels powerful because it arrives *after* meaning has already formed.

That timing matters.

Under DELTA, language is not where meaning originates, and it is not where truth resides. Language is an interface—a negotiated protocol that allows one organized resolution system to perturb another using compressed symbols instead of shared experience.

Words do not carry reality. They carry alignment hints.

When a system compresses stabilized alignment into meaning, that compression is already lossy. When it is translated into language, it is compressed again. What moves from speaker to listener is not experience, not understanding, not truth—but a trigger. The listener must re-resolve internally for anything meaningful to occur.

This explains why language works remarkably well—and fails constantly.

Language works when internal alignment structures are similar enough that compressed cues land near compatible pathways. A few words can reorganize coherence rapidly. Insight feels immediate. Communication feels "clear."

Language fails when alignment architectures differ. The same words inject delta into incompatible loops. Misalignment increases instead of resolving. Confusion, conflict, or ideology follow—not because words were wrong, but because interfaces were mismatched.

This also explains why agreement is not truth.

Agreement is interface compatibility, not reality alignment. Groups synchronize language long before they synchronize understanding. Shared vocabulary can mask deep misalignment. This is why slogans spread faster than insight, and why technical language can simulate knowledge without producing it.

Language optimizes for transfer efficiency, not fidelity.

It trades depth for speed. It allows coordination without shared experience. This makes civilization possible—but it also makes distortion inevitable. The more abstract the language, the further it drifts from lived alignment.

This reframing dissolves a common frustration: the feeling that words fail to capture what "really happened."

They are supposed to.

Language was never meant to preserve experience. It was meant to perturb alignment just enough to allow another system to reconstruct something usable. Precision is achieved not by better words, but by shared context, timing, and coherence.

This is also why silence can communicate more than speech. When systems share alignment already, words add noise. Silence preserves coherence. Language becomes necessary only when alignment must be rebuilt.

Seen this way, language is not a mirror of reality. It is a tool for coordination under compression.

Mistaking language for truth leads to dogma. Mistaking silence for ignorance leads to arrogance. Both errors arise from forgetting what language is for.

Understanding language as interface completes the picture of mind under DELTA. The observer resolves. Awareness registers. Meaning compresses. Memory stabilizes. Language exchanges alignment imperfectly.

Nothing mystical was required.

What remains is the question people care about most: choice, agency, and freedom. If mind is embedded in resolution, what does it mean to choose?

That question does not break DELTA. It sharpens it.

And that is where we turn next.

13.6 — Free Will Revisited

Free will is usually framed as a yes–no question: either we are free, or everything is determined. Under DELTA, this framing collapses. It assumes freedom must mean independence from physics. That assumption is the error.

Freedom is not independence. Freedom is bandwidth.

A system has freedom to the extent that it can redirect delta before resolution completes. Where alignment is rigid, resolution runs downhill and choice disappears. Where alignment capacity is wide, resolution can be rerouted, delayed, or reshaped. What we experience as choice is this window of redirection.

Choice is not creation of outcomes. It is navigation among constrained possibilities.

When a stimulus arrives, delta is injected. In a low-bandwidth system—reflexive, overloaded, fragmented—resolution proceeds immediately along the strongest pathway. Reaction occurs. No choice is felt because none is physically available.

In a high-bandwidth system, internal coherence holds long enough to allow comparison, memory access, and alternative routing. Resolution pauses—not because of a ghostly will, but because alignment capacity permits delay. That delay is the space in which choice appears.

This explains why freedom varies.

You are less free when exhausted, stressed, frightened, intoxicated, or traumatized. Delta overwhelms ATT. Resolution collapses into reaction. You are more free when rested, calm, coherent, and supported. Alignment holds. Redirection becomes possible.

Freedom is therefore state-dependent, not absolute.

This reframing dissolves the guilt trap often attached to free will. If choice were absolute, failure would be moral defect. Under DELTA, failure is often bandwidth exhaustion. Responsibility does not vanish—but it shifts from blame to coherence management.

Responsibility is not about control. It is about maintaining the conditions under which redirection remains possible.

This also explains why training, discipline, and ethics matter physically. They do not impose morality from outside. They expand alignment bandwidth—through habit, reflection, pacing, and structure—so that delta can be handled without collapse.

Free will is not a miracle inside matter. It is what happens when matter is organized well enough to wait.

Seen this way, determinism and freedom are not opposites. Determinism describes what happens when bandwidth is exhausted. Freedom describes what becomes possible when bandwidth is preserved.

The mind is not free because it floats above physics. It is free when physics gives it room.

This understanding clears the final conceptual ground. There is no need to defend consciousness as magical or dismiss it as illusion. Consciousness is simply awareness sustained across sufficient bandwidth.

That is the final clarification of this chapter—and the bridge to its close.

13.7 — Consciousness Without Mysticism

Consciousness has been inflated into a riddle by treating it as a thing.

A substance. A light that turns on. A property added to matter when it becomes complicated enough.

DELTA dissolves this confusion by refusing to add anything at all.

Consciousness is not an entity. It is a bandwidth condition.

When awareness—internal coherence—can be sustained across time, across loops, and across changing deltas, the system experiences continuity of experience. That sustained awareness is what we label consciousness. Nothing extra appears. Nothing descends. Alignment simply holds long enough to remain visible to itself.

This immediately explains why consciousness fades.

Sleep, anesthesia, concussion, deep exhaustion—all reduce alignment bandwidth. Internal loops desynchronize. Awareness fragments. The system continues to function locally, but self-registration collapses. Consciousness is not "turned off." It becomes physically unavailable.

It also explains why altered states feel profound.

Meditation, flow, psychedelics, trauma, prayer—very different triggers, same mechanism. Delta is redistributed. Some alignment pathways quiet. Others synchronize intensely. Bandwidth narrows in some dimensions and widens in others. Awareness becomes unusually focused or diffuse. The experience feels "bigger" or "deeper" not because reality changed, but because coherence geometry shifted.

Nothing supernatural is required.

This framing also removes the temptation to place consciousness outside evolution or physics. Consciousness does not float above life. It emerges naturally when life's internal loops become coherent enough to reference themselves continuously.

There is no observer behind the observer. No witness watching awareness.

That infinite regress is an illusion created by language.

There is only resolution, alignment, registration—and continuity when conditions allow it.

This also clarifies why consciousness is fragile. It is expensive. It requires sustained coherence. Many organisms do not need it. Some lose it temporarily to survive. The universe does not optimize for consciousness. It merely permits it under rare conditions, just as it permits life.

Consciousness, then, is not the crown of creation. It is a narrow operating mode of organized matter.

Seen this way, debates about whether machines "will be conscious" lose their drama. The question is not silicon versus carbon. The question is whether a system can sustain awareness bandwidth under continuous delta. If it can, consciousness follows. If it cannot, it will not—regardless of intelligence.

This closes the loop that began at the start of the book.

Delta drives all dynamism. Alignment governs how it resolves. Life loops resolution. Mind registers it. Consciousness sustains that registration.

Nothing mystical was added at any stage.

What remains is integration—how these insights reshape responsibility, ethics, civilization, and survival. That work lies ahead.

For now, one conclusion stands firm:

Consciousness is not a mystery of matter. It is matter organized well enough to keep noticing itself.

From Understanding to Responsibility

With mind, awareness, and meaning placed back inside physics, the last illusion quietly dissolves: that understanding is neutral.

It is not.

To understand how delta resolves is to realize that every system participates in shaping its own outcomes, whether it knows it or not. Living systems do not float through reality. They negotiate it continuously—by pacing delta, preserving alignment, and protecting bandwidth.

The mind, now fully demystified, carries a new weight.

Awareness is not a privilege. It is a burden of coherence.

To be aware is to hold resolution open long enough to redirect it. To sustain consciousness is to remain responsible for how delta moves through you. Ignorance is not just lack of knowledge; it is collapsed bandwidth. Wisdom is not accumulation of facts; it is alignment management.

Meaning, memory, language, and choice are no longer abstract concepts. They are tools—tools that can preserve coherence or shatter it, depending on how they are used. Systems that misuse these tools accelerate dissipation. Systems that respect alignment survive longer, adapt better, and harm less.

This realization carries consequences beyond the individual.

Civilizations are alignment systems. Institutions are delta-routing architectures. Ethics are coherence-preserving constraints.

When delta exceeds alignment capacity—socially, technologically, psychologically—collapse follows. Not because of evil, but because resolution outran structure.

The remaining parts of this book turn outward. They ask what DELTA demands of intelligence once it understands itself. How responsibility scales. How power, technology, and culture can either expand bandwidth or annihilate it.

We do not move forward by controlling the world. We move forward by learning how not to exceed ourselves.

From here on, the question is no longer *what is reality?* It is what must an aware system do to remain coherent inside it?

That is where understanding becomes obligation.

The observer is not outside reality. The observer is reality resolving with memory.

Interlude — Putting the Observer Back Where It Belongs

Somewhere along the way, the observer was promoted too far.

What began as a practical acknowledgment—that measurement involves interaction—slowly inflated into a worldview where reality waits to be seen, collapses when looked at, and depends on awareness to exist at all. This was not malice. It was confusion born from mixing experience with existence.

DELTA does not permit that confusion.

Nature does not pause for observers. Delta resolves whether or not it is registered. Alignment forms, breaks, and reforms long before any system becomes capable of noticing it. Rivers flowed before eyes evolved. Stars burned before minds arose. Life itself stabilized long before awareness reflected on it.

The observer did not create reality. The observer entered it.

What awareness adds is not authority, but access. An observer is a resolution system that can *participate knowingly* in dynamics already underway. Observation does not grant power over existence; it grants exposure to consequence.

This interlude marks a necessary correction.

Until now, we have traced a clean arc:

delta → resolution → life → mind → meaning.

At no point did reality require permission to proceed.

The danger begins when awareness mistakes visibility for control. When registration is confused with authorship. When access is mistaken for dominance. That confusion has shaped how science is narrated, how technology is wielded, and how power justifies itself.

Putting the observer back in place is not a downgrade. It is a stabilization.

An observer is not a god, not a puppet master, not a passive witness. An observer is an aligned participant—capable of redirecting delta locally, but never exempt from global consequence.

This realization is not philosophical. It is operational.

If observation does not create reality, then responsibility replaces entitlement. If manipulation does not equal domination, then ethics becomes structural, not moral. If power is misalignment management, then excess power is instability, not strength.

The next part begins there.

Not with denial of human capability, but with its proper framing.

Once the observer is put back in place, we can finally ask the right questions—not about how to control reality, but about how to interact with it without collapsing coherence.

That is the task of next part.

Part V — The Observer Put Back In Place

This part corrects a critical imbalance that modern thinking has quietly normalized: the elevation of the observer from participant to authority.

This part reasserts a simple but destabilizing truth—reality does not require observation to exist, and systems do not grant control simply because they are understood. Observation is access, not authorship. Manipulation is navigation, not domination.

By grounding the observer back inside dynamism, this part dismantles observer-centric models that confuse experience with existence and replaces them with a physically honest view of participation and consequence. Technology, power, and ethics are reframed as methods of routing delta within constraints—not as tools for overriding them.

This section shows why control always fails when it ignores alignment capacity, why domination accelerates collapse, and why ethics is not moral decoration but structural necessity. Once the observer is placed correctly, responsibility replaces illusion—and interaction becomes sustainable.

This part does not reduce human agency. It stabilizes it.

Chapter 14 — Observation Does Not Create Reality

Reality Was Busy Before We Arrived

There is a quiet arrogance hidden in the way modern language talks about observation.

We say reality "collapses" when measured. We imply that the universe waits, undecided, until awareness arrives. We flirt with the idea that consciousness brings the world into being. These ideas feel profound because they elevate experience—but they do so by misplacing it.

Reality did not begin when it was noticed.

Stars burned for billions of years before eyes evolved. Mountains rose, eroded, and vanished without witnesses. Life stabilized its internal loops long before any system became capable of reflecting on them. Delta resolved relentlessly, patiently, without needing to be acknowledged.

Observation came late.

The confusion began innocently. Measurement in physics involves interaction. Interaction affects outcomes. Somewhere along that chain, *affects* quietly became *creates*. A practical caveat hardened into an ontological claim. The observer was promoted from participant to author.

DELTA refuses that promotion.

Observation does not grant authority over existence. It grants exposure to consequence. To observe is to insert oneself into a process already underway—not to summon that process into being.

This distinction matters because observer-centric thinking is unstable. It cannot scale. If reality depends on being observed, what observed the universe before observers existed? If awareness creates outcomes, why does ignorance not protect us from consequence?

Experience is real. Existence is independent.

Conflating the two flatters the mind but fractures coherence. It turns knowledge into entitlement, curiosity into control, and understanding into domination. Entire technological and philosophical missteps trace back to this single confusion.

Putting the observer back in place is not a denial of meaning. It is a rescue of it.

This chapter begins by doing something deceptively simple: restoring chronology. Nature first. Observers later. Reality continuous throughout.

Once that order is respected, many paradoxes dissolve on their own—and responsibility re-enters the picture where illusion once stood.

Reality was busy before we arrived. It will remain busy long after. The task of observation is not to create it, but to meet it honestly.

14.1 — Nature Before Observers

Long before there were eyes to see, ears to hear, or minds to wonder, nature was already at work.

Galaxies formed, collided, and dispersed. Stars ignited, exhausted their fuel, and collapsed. Planets cooled, cracked, and reshaped themselves. Chemical gradients organized, broke apart, and reorganized again. None of this waited for observation. None of it paused for interpretation.

From a DELTA perspective, this is not a philosophical claim. It is a physical one.

Delta resolves wherever conditions permit it. Alignment forms and breaks according to structure and timing, not awareness. The universe does not require registration to proceed because registration is not a prerequisite for resolution.

This is where observer-centric models quietly fail their first test: chronology.

If reality depended on observers, then pre-biological history would be unintelligible. Geological time would be meaningless. Cosmology would collapse into self-reference. Yet the evidence is overwhelming—not just inferential, but causal. Radioactive decay proceeded. Light traveled. Fields interacted. Structures evolved. All without witnesses.

Existence precedes experience.

What observers encounter is not creation-in-progress, but ongoing consequence. By the time a system is observed, delta has already resolved through countless stages. Observation samples a slice of an unfolding process; it does not initiate it.

This also explains why nature remains stubbornly indifferent to belief, attention, or neglect. Gravity does not weaken when ignored. Disease does not disappear when misunderstood. Climate does not respond to denial. These are not moral statements—they are alignment facts. Delta resolves regardless of who is watching.

DELTA makes this unavoidable: resolution is self-sufficient.

Observers arise *within* nature because certain systems evolve the capacity to loop resolution internally and reference it. Awareness is an emergent capability, not a cosmic trigger. It grants access to what is happening, not permission for it to happen.

Putting this plainly: the universe does not need to be seen to exist. It needs to be structured to resolve.

Once this order is restored—nature first, observers later—many seductive ideas lose their footing. Consciousness stops being cosmic glue. Observation stops being magical. What remains is something quieter and more demanding: participation without primacy.

With this grounding in place, we can now examine why observer-centric models persist despite this mismatch—and why they fail when pushed beyond metaphor.

That is where we turn next.

14.2 — Why Observer-Centric Models Fail

Observer-centric models fail not because they are useless, but because they overextend a local truth into a global claim.

At small scales, especially in quantum experiments, measurement clearly affects outcomes. Interaction changes states. Coupling alters resolution paths. This is real, measurable, and unavoidable. The mistake begins when *interaction* is quietly promoted to *creation*.

That promotion is a category error.

An observer interacting with a system injects delta. The system responds by resolving that delta within its alignment constraints. Outcomes change—not because reality was undecided, but because the resolution pathway was perturbed. Observer-centric models stop at this point and leap to ontology: *the observer caused reality to exist in that form*.

This leap does not survive scale.

If observation creates reality, then unobserved systems should lack defined behavior. Yet stars evolve predictably, radioactive elements decay statistically, tectonic plates move relentlessly, and biological evolution proceeds without supervision. The universe does not become vague when unattended. It becomes inaccessible, not indeterminate.

Observer-centric interpretations also collapse under consistency. Different observers with different frames, instruments, and timing still converge on stable, reproducible structures. If reality were observer-authored, convergence would be miraculous. Instead, convergence is expected—because existence constrains experience, not the other way around.

Another failure appears in causality. Observer-centric thinking subtly reverses cause and effect. Measurement outcomes are treated as if awareness selects them, when in fact awareness arrives *after* resolution has already been forced by interaction. The observer encounters the consequence of delta injection, then retroactively narrates it as authorship.

This inversion flatters the mind but destabilizes explanation.

It also leads to intellectual inflation. If observation creates reality, then understanding implies control, and knowledge implies exemption from consequence. This belief quietly underpins technological overreach, ethical shortcuts, and the fantasy that insight alone grants mastery.

DELTA exposes the flaw cleanly.

Observation changes systems because it participates in them. It does not create systems because systems were already resolving.

Observer-centric models work descriptively at the interface—where measurement meets outcome—but fail ontologically when they claim primacy. They confuse access with authority.

Once this distinction is restored, quantum measurement loses its metaphysical drama. It becomes what it always was: a constrained interaction between resolution systems, governed by alignment bandwidth and ATT.

Reality does not collapse because it is seen. It resolves because delta was introduced.

Recognizing this clears the path for the next necessary distinction: between what is experienced internally and what exists externally. Confusing those two is where observer-centrism does its deepest damage.

That distinction is our next step.

14.3 — Experience vs Existence

The deepest confusion in observer-centric thinking is not about physics. It is about where experience ends and existence begins.

Experience is undeniable. Whatever you feel, perceive, or register is real *as experience*. Pain hurts. Color appears. Thought unfolds. DELTA does not question this. But experience being real does not make it ontologically authoritative.

Existence does not wait to be experienced.

From a DELTA perspective, experience is internal resolution—delta resolving within an organized system that can register alignment changes. Existence is external resolution—delta resolving in the world regardless of registration. The two are linked, but they are not the same.

Confusing them produces paradox.

When experience is mistaken for existence, perception becomes proof. "I saw it" becomes "it is." When existence is mistaken for experience, absence of perception becomes absence of reality. Both errors arise from collapsing two different resolution domains into one.

Experience answers the question: *what happened to the observer?* Existence answers the question: *what happened, period?*

This distinction explains why reality routinely violates expectation. Earthquakes occur without warning. Diseases progress unnoticed. Systems fail quietly before anyone realizes it. Existence resolves delta on its own timetable, not on the observer's awareness cycle.

It also explains why subjective certainty is unreliable. Strong experience—fear, awe, conviction—feels authoritative because internal alignment is intense. But intensity of internal resolution does not grant control over external resolution. A belief can feel absolute while being physically irrelevant.

DELTA makes this blunt: internal coherence does not rewrite external dynamics.

This is why observer-centric philosophies drift toward solipsism. If experience is treated as primary, the world becomes a projection. But projections do not push back reliably. Reality does.

Existence constrains experience relentlessly. Gravity does not negotiate. Fire does not reinterpret intention. Time does not slow because awareness wishes it to. These constraints are not cruel; they are structural.

Putting experience and existence in their proper places does not diminish meaning. It stabilizes it.

Experience matters because it guides how an observer navigates existence. It does not matter because it authors existence. When this boundary is respected, learning becomes possible. When it is violated, ideology replaces adaptation.

Experience is how reality arrives inside a system. Existence is what reality does regardless.

Once this separation is restored, the role of alignment becomes clear. Awareness does not grant power over being. It grants access to consequence—the ability to sense, anticipate, and respond.

That access, properly understood, is not authority. It is responsibility.

And that brings us to the next clarification: alignment as access, not dominance.

14.4 — Measurement as Alignment Injection

Measurement feels passive because its effects are often subtle. A glance, a reading, a click of a detector—it does not *feel* like an intervention. But physically, measurement is never neutral.

Measurement is alignment injection.

Whenever an observer measures a system, they introduce coupling: energy exchange, constraint, boundary definition, timing synchronization. Even the gentlest measurement inserts delta into a system that was already resolving in its own way. The system then responds—not to being *seen*, but to being perturbed.

This distinction matters.

Measurement does not ask reality a question and wait for an answer. It joins the system's resolution pathway and alters the available routes. Outcomes change because the landscape of alignment has changed—not because existence was undecided.

At quantum scales, alignment bandwidth is narrow. Even minimal coupling overwhelms continuity. Resolution jumps. The observer's presence becomes dominant not because awareness is powerful, but because the system has little room to absorb disturbance. Measurement outcomes differ sharply because the injected delta is proportionally large.

At macroscopic scales, alignment bandwidth is wide. The same injection is absorbed smoothly. Measurement barely registers. Outcomes appear stable. The difference is not philosophical—it is capacity.

Seen through DELTA, measurement is neither revelation nor creation. It is participation with consequences.

This reframing dissolves the drama around "collapse." What collapses is not reality; it is a description that assumed isolation. Once interaction occurs, isolation is gone. The system resolves under new constraints, and our model updates accordingly.

Importantly, this also explains why different measurement setups yield different results. Change the apparatus, and you change the alignment injection—geometry, timing, coupling strength. The system responds differently because it is resolving a different delta. Nothing mystical selects outcomes. The resolution follows structure.

This perspective also clarifies limits.

There is no perfectly non-invasive measurement. There is only measurement gentle enough that its delta falls within the system's

ATT. Where that is possible, continuity survives. Where it is not, discontinuity appears. The boundary is physical, not epistemic.

Measurement, then, is a responsibility.

To measure is to alter. To alter is to participate. To participate is to bear consequence.

Understanding measurement as alignment injection puts the observer back where it belongs—not outside reality issuing commands, but inside reality sharing its constraints.

With this correction in place, the final step becomes obvious. If observation grants access but not authorship, then alignment itself must be reinterpreted—not as a lever of control, but as a corridor of permission.

That is the focus of the next section.

14.5 — Alignment as Access, Not Authority

Alignment is often mistaken for power.

When systems align well—when understanding deepens, predictions improve, and interventions become effective—it is tempting to believe that alignment grants authority over reality. That temptation sits at the root of observer-centric thinking, technological overconfidence, and ethical shortcuts.

DELTA makes a harder claim: alignment does not confer control. It confers access.

To be aligned with a system is to be exposed to its consequences more clearly, not to be exempt from them. Alignment opens corridors through which delta can be sensed, anticipated, and navigated. It does not rewrite the terrain.

This distinction is subtle but decisive.

Authority implies command: the right to decide outcomes. Access implies participation: the obligation to respond accurately.

When alignment improves, actions become more effective because they are better timed and better placed—not because the system has become submissive. The river does not obey the engineer; the engineer learns where the river will flow if guided gently. The body does not obey medicine; medicine succeeds only when it respects physiology's alignment constraints.

Alignment increases sensitivity, not sovereignty.

This is why deeper understanding often increases vulnerability. The more access a system has, the more it can be harmed by misalignment. Knowledge amplifies consequence. Technology magnifies error. Insight removes the protection of ignorance.

DELTA reframes this as a physical truth, not a moral warning.

As alignment improves, delta injected by action propagates farther and faster. Small mistakes ripple widely. Systems become efficient—and therefore fragile. What looks like mastery is actually reduced tolerance for error.

This is why domination always fails eventually.

Domination treats alignment as authority. It forces delta beyond ATT, extracts short-term control, and ignores long-term coherence. The system responds predictably: misalignment accumulates, feedback loops collapse, and resolution accelerates toward breakdown. Collapse is not rebellion. It is physics finishing the sentence.

Ethics, under DELTA, emerges naturally here.

Ethics is not restraint imposed from outside. It is structural respect for alignment limits.

To act ethically is to navigate delta without exceeding coherence capacity—of oneself, of others, of systems that cannot speak but still resolve. To act unethically is not to be evil; it is to be misaligned at scale.

This reframing removes moral theater and replaces it with responsibility grounded in consequence. Awareness does not grant the right to dominate. It grants the duty to handle access carefully.

Alignment, properly understood, humbles the observer.

You do not command reality because you understand it. You are entrusted with narrow corridors of participation.

Once this is accepted, the observer is finally back in place: powerful locally, constrained globally, responsible everywhere. From here, the final synthesis becomes unavoidable—how manipulation can occur without domination, and how power can be exercised without collapse.

That synthesis completes this part.

14.6 — The Observer Put Back in Place

Putting the observer back in place is not an act of demotion. It is an act of stabilization.

For too long, the observer has been oscillating between two extremes: either irrelevant (a passive recorder of an objective world) or omnipotent (a consciousness that collapses reality into being). Both positions are physically untenable. One denies interaction; the other inflates it.

DELTA allows neither illusion.

The observer is an embedded resolution system—no more, no less.

Embedded means constrained. The observer operates within alignment limits, ATT thresholds, and structural dependencies. Awareness does not free the observer from physics; it binds the observer more tightly to consequence. Every act of observation, interpretation, and intervention injects delta that must resolve somewhere.

This is why humility is not philosophical modesty—it is physical accuracy.

An observer cannot step outside resolution. There is no privileged vantage point where consequences do not return. Knowledge increases reach, but reach increases load. Bandwidth is not infinite. Every gain in access narrows tolerance for error.

Putting the observer back in place also clarifies responsibility.

Responsibility is not moral burden imposed by society. It is what naturally follows when access expands.

An unaware system can only react. An aware system can redirect. That redirection capability makes the system accountable—not by law or doctrine, but by structure. If delta is misrouted, the observer absorbs the cost internally or externally. Reality does not negotiate.

This reframing dissolves the fantasy of neutrality.

There is no neutral observation. There is no harmless insight. There is only participation with varying degrees of foresight.

The observer's true position, then, is narrow and demanding: capable of local influence, incapable of global exemption. Able to shape pathways, unable to rewrite rules. Free within bandwidth, bound by alignment.

When this position is respected, systems remain coherent. When it is violated—when observers mistake access for authority—collapse follows, quietly and inevitably.

Putting the observer back in place does not diminish science, meaning, or agency. It rescues them from overreach. It restores scale, proportion, and consequence.

With the observer stabilized, the final chapter of this part can proceed honestly. If observation does not create reality, and alignment does not grant authority, then what remains of power?

What remains is navigation without domination.

That question closes this part—and prepares the ground or everything that follows.

From Access to Accountability

With the observer finally returned to its proper place, one truth becomes unavoidable: access carries weight.

To see more clearly is not to rise above consequence—it is to become more tightly bound to it. Awareness widens corridors of interaction, not escape routes. Alignment sharpens sensitivity, not immunity. Every gain in understanding increases the precision with which delta propagates—and therefore increases the cost of error.

This is where modern thinking often stumbles.

We celebrate insight as if it were ownership. We treat knowledge as entitlement. We assume that because a system can be understood, it can be overridden. DELTA exposes the flaw: understanding expands participation, not authority.

From this point on, power must be redefined.

Power is not the ability to force outcomes. It is the capacity to route delta without exceeding coherence limits.

When that capacity is respected, systems adapt and endure. When it is abused, domination masquerades as control until alignment collapses. Collapse is not resistance—it is delayed consequence.

This closing does not preach restraint. It states a physical condition.

Technology, institutions, and intelligence cannot escape alignment laws. The more they amplify delta, the more precisely they must manage it. Ethics ceases to be optional the moment access expands beyond local scale.

With observation stripped of authorship and alignment stripped of authority, the path forward becomes starkly practical. The question is no longer whether we *can* intervene—but whether we can bear what our interventions release.

The next part moves from correction to synthesis. It asks how manipulation can occur without domination, how intelligence can act without collapse, and how responsibility can scale without paralyzing progress.

From here on, understanding is no longer enough.

What matters is how we move delta once we know where it goes.

Reality does not require observation. Observation requires reality.

Chapter 15 — Manipulation Without Domination

Power That Does Not Break What It Touches

Power is usually celebrated for its effects, not its aftermath. When a system bends, yields, or accelerates under intervention, we call it success. When outcomes align with intent, we call it control. Only later—often much later—do we notice the fractures: fatigue, resistance, brittleness, collapse. By then, power has already mistaken compliance for stability.

This chapter begins with a correction.

Manipulation is not the problem. Domination is.

Every interaction with reality is a form of manipulation. Tools manipulate materials. Language manipulates attention. Technology manipulates scale. Even restraint manipulates outcomes by choosing *not* to inject delta. The question is never whether manipulation occurs, but how delta is routed once it is introduced.

Domination fails because it treats alignment as something to be enforced rather than navigated. It assumes that if enough force, incentive, or pressure is applied, coherence will follow. It does—briefly. Then misalignment accumulates invisibly, feedback loops weaken, and the system begins to decay from within.

Power feels strongest right before it becomes most fragile.

DELTA reframes power in a way that removes both moral drama and technological arrogance. Power is not the ability to override systems. It is the ability to intervene without destroying the system's capacity to resolve future delta.

This distinction is subtle but decisive.

A dam that channels a river without choking it exercises power. A dam that blocks flow entirely invites rupture.

A policy that nudges behavior while preserving agency stabilizes society. A policy that enforces uniformity breeds silent divergence.

A technology that amplifies capability while respecting alignment limits expands possibility. A technology that ignores those limits accelerates collapse.

In each case, the difference is not intention or intelligence. It is alignment literacy.

This chapter does not argue for restraint as virtue. It argues for restraint as physics. Systems that are pushed beyond their resolution capacity do not protest. They fail. Often politely. Often later.

The challenge of the modern world is not insufficient power. It is power without placement—intervention divorced from consequence, scale divorced from responsibility.

Manipulation without domination is not idealism. It is the only form of power that survives its own success.

From here, the chapter moves into operational ground: technology, control, ethics, and leadership—not as moral categories, but as ways delta is handled when awareness is present.

The question is no longer whether we can shape outcomes. It is whether we can do so without breaking the system that must carry them next.

15.1 — Technology as Delta Navigation

Technology is often framed as power over nature. That framing is inaccurate—and dangerous.

Technology does not create new laws, new forces, or new freedoms. It reshapes how delta enters and resolves within existing systems. Every tool, from a stone blade to an algorithm, is a mechanism for navigating misalignment—not eliminating it.

Seen through DELTA, technology is structured amplification.

A lever does not defeat gravity; it redistributes load. An engine does not create motion; it redirects energy gradients. A network does not invent intelligence; it accelerates alignment exchange.

In each case, technology widens the corridor through which delta can flow. It increases scale, speed, reach, or precision—but it never suspends alignment limits. It merely moves the point at which those limits are felt.

This is why technology always carries delayed consequences.

When delta is amplified faster than a system's ATT can absorb, misalignment does not vanish. It migrates. It accumulates in places that are harder to see: ecological strain, psychological overload, social fragmentation, systemic fragility. Early success hides future cost because resolution has been deferred, not completed.

Technology succeeds when it behaves like navigation, not imposition.

Navigation respects terrain. It adapts to gradients. It works *with* structure rather than against it. Good technologies sense feedback early, distribute load, and allow local correction. They preserve the system's capacity to resolve future delta.

Bad technologies pursue straight lines.

They optimize for immediate output, suppress feedback, and centralize control. They force alignment through pressure rather than pacing. Such systems appear efficient—until they exhaust coherence. Failure arrives not as resistance, but as brittleness.

This reframing explains a recurring historical pattern: every major technological leap solves one alignment problem while creating a larger one downstream.

Fire freed energy but demanded social control. Agriculture stabilized food but introduced hierarchy and disease. Industry multiplied output but strained ecosystems and minds. Digital systems expanded cognition but fractured attention and trust.

None of these failures are moral. They are navigation errors.

Technology is not neutral because delta is not neutral. Amplified misalignment travels farther. The more powerful the tool, the narrower the margin for error. This is why advanced technology demands greater alignment literacy, not just greater intelligence.

DELTA makes this unavoidable: power grows faster than tolerance.

The task, then, is not to abandon technology or worship it. It is to design technologies that route delta gently, preserve feedback, and respect the resolution capacity of the systems they touch.

Technology does not grant dominance over reality. It grants access to deeper consequence.

Understanding this prepares us for the next uncomfortable truth: attempts to convert navigation into control always fail—not immediately, but inevitably.

That failure, and why it repeats so reliably, is where we turn next.

15.2 — Why Control Always Fails Eventually

Control promises certainty. That is why it is so seductive.

When outcomes become predictable under pressure—when systems comply, behaviors align, and variability shrinks—it feels like mastery. But what control actually does is force alignment beyond a system's natural resolution bandwidth. The appearance of stability is purchased by suppressing feedback, not by resolving misalignment.

DELTA explains why this cannot last.

Every system has an Alignment Threshold Time (ATT)—a finite rate at which it can absorb, distribute, and resolve injected delta. Control ignores this limit. It accelerates resolution artificially, compressing time, narrowing options, and concentrating stress. The system responds by deferring resolution internally.

Deferred resolution is not eliminated. It accumulates.

This accumulation is invisible at first because controlled systems are quiet. Noise is suppressed. Variance is punished. Signals flatten. What disappears is not misalignment, but its early warnings. By the time symptoms reappear, the system's capacity to self-correct has already been eroded.

This is why controlled systems rarely fail under peak stress. They fail after prolonged stability.

History shows this pattern repeatedly. Economics collapse after long booms. Organizations implode after years of flawless performance. Individuals burn out not during chaos, but after sustained over-optimization. Control creates an illusion of order while quietly draining resilience.

The deeper flaw in control is structural.

Control assumes that systems can be held in preferred states indefinitely. But preferred states are alignment snapshots, not resting points. Delta never stops resolving. When resolution pathways are blocked, energy is forced into secondary channels—often destructive ones.

This is why tighter control requires ever-increasing effort. Each layer of enforcement adds new misalignment that must itself be controlled. Eventually, the system becomes more about maintaining control than performing its function.

At that point, collapse is efficient.

DELTA reframes collapse not as failure of authority, but as completion of deferred resolution. The system releases stored misalignment in a rapid, often catastrophic adjustment. What looks like sudden breakdown is actually long-delayed honesty.

Control fails because it treats living, adaptive systems as static machines. It enforces outcomes without respecting timing, bandwidth, or feedback. The result is temporary order followed by irreversible damage.

Understanding this shifts the question.

The alternative to control is not chaos. It is navigation with feedback.

The task is not to freeze systems into compliance, but to guide delta through channels that preserve adaptability. That requires responsibility rather than dominance, humility rather than force.

With control stripped of its false promise, the next issue becomes unavoidable: if domination fails structurally, what replaces it? The answer is not morality—it is ethics grounded in delta responsibility.

That is where we go next.

15.3 — Ethics as Delta Responsibility

Ethics is usually treated as a moral overlay—values imposed on action after the fact. Under DELTA, ethics precedes morality. It is not about intention, virtue, or goodness. It is about how delta is handled once awareness exists.

Ethics is responsibility for routed misalignment.

The moment a system gains the ability to intervene—to inject delta deliberately rather than reactively—it inherits a new obligation. Not a social obligation. A physical one. Delta introduced intentionally travels farther, faster, and with sharper consequences than delta introduced accidentally. Ethics emerges because awareness magnifies reach.

This is why scale transforms neutral acts into harmful ones.

At small scales, misalignment is absorbed locally. Errors self-correct. At large scales, the same actions overwhelm alignment capacity elsewhere. A decision that is harmless in isolation becomes destructive when amplified by technology, authority, or repetition. Ethics is the discipline of anticipating where amplified delta will land.

DELTA strips ethics of sentimentality.

Ethical action is not about being kind. It is about not exceeding coherence limits—of people, ecosystems, institutions, and oneself.

Unethical action is not wickedness. It is misalignment at scale.

This reframing explains why good intentions routinely produce bad outcomes. Intent does not alter physics. A well-meaning intervention that ignores ATT will still fracture systems. Conversely, a restrained intervention that respects alignment may appear slow or conservative, yet preserve long-term viability.

Ethics, then, is not restraint for its own sake. It is precision under amplification.

As access grows, tolerance for error shrinks. This is why advanced systems require stronger ethical constraints, not weaker ones. The more powerful the tool, the narrower the safe corridor. Ethics is the map of that corridor.

This also clarifies why ethical systems decay when treated as ideology. Rules detached from consequence become brittle. They enforce compliance rather than preserve coherence. True ethics is adaptive. It evolves with scale, technology, and context because delta pathways change.

Responsibility, in this framework, is not guilt. It is load-bearing awareness.

To know more is to carry more. To act widely is to answer for wider ripples. Ethics is simply the name we give to this physical truth once it enters human language.

With ethics grounded in responsibility rather than morality, the final step becomes clear. Power itself must be redefined—not as dominance or control, but as the capacity to manage misalignment without collapse.

That redefinition is the focus of the next section.

15.4 — Power as Misalignment Management

Power is most often measured by what it can impose. DELTA measures power by what it can absorb.

Every system—natural, social, technological, biological—exists in a state of continuous misalignment. Delta never disappears; it shifts, redistributes, and resolves. Power, in this context, is not the ability to eliminate misalignment, but the capacity to manage it without losing coherence.

This reframing changes everything.

A powerful system is not one that enforces uniformity. It is one that can tolerate variance without fragmentation. Uniform systems look stable, but they are brittle. Diverse systems appear noisy, but they are resilient. Power lies in holding divergence without collapse.

This is why resilient organizations decentralize. Why ecosystems thrive on diversity. Why bodies maintain health through redundancy. Each distributes delta across many pathways, preventing overload in any single channel. Power here is architectural, not coercive.

Misalignment management requires three capabilities: sensing, buffering, and routing.

Sensing means detecting divergence early—before resolution becomes violent. Buffering means absorbing excess delta temporarily without transferring it destructively. Routing means guiding misalignment into channels where it can resolve productively.

Systems that lack these capabilities resort to force. Force feels decisive, but it bypasses all three functions. It suppresses signals, removes buffers, and blocks routing. The result is apparent order followed by sudden failure.

This is why leadership based on pressure scales poorly. It may work locally or briefly, but it cannot handle complexity. As scale increases, misalignment multiplies faster than enforcement capacity. Power then becomes self-consuming.

DELTA exposes a quiet truth: power that cannot process misalignment will be consumed by it.

This applies equally to individuals. Emotional power is the ability to hold internal conflict without impulsive action. Intellectual power is the ability to entertain contradiction without collapse. Moral power is the ability to remain coherent under ambiguity. In each case, dominance over others is irrelevant; self-regulation is central.

At societal scales, power is stewardship.

Stewardship is often misunderstood as weakness. In reality, it is the most demanding form of power. It requires resisting the temptation to simplify, suppress, or accelerate beyond capacity. It accepts that misalignment is permanent and chooses to work with it intelligently.

Seen this way, power is no longer a trophy. It is a continuous operational burden.

Those who seek power to feel strong are unfit for it. Those who accept power as responsibility may be capable of carrying it.

With power redefined as misalignment management, the final illusion to be dismantled becomes obvious: the fantasy that systems can be perfectly optimized.

That illusion—and why it destroys coherence—is what we confront next.

15.5 — The Illusion of Total Optimization

Optimization feels intelligent because it promises clarity.

Choose a goal. Define a metric. Eliminate everything that does not serve it. When numbers improve, confidence follows. This logic works beautifully in closed, simple systems. It fails catastrophically in living, adaptive ones.

DELTA exposes why.

Total optimization assumes that a system can be reduced to a single dominant alignment axis. Reality never complies. Every real system resolves multiple deltas simultaneously, across different timescales and layers. Optimizing one channel inevitably starves others.

What looks like efficiency is often misalignment deferred elsewhere.

When productivity is maximized, resilience is usually sacrificed.

When speed is optimized, understanding thins. When profit is optimized, trust erodes. When control is optimized, adaptability dies.

These trade-offs are not managerial failures. They are structural consequences of narrowing resolution pathways.

Total optimization collapses systems because it removes slack.

Slack is not waste. Slack is alignment buffer. It is what allows systems to absorb unexpected delta without breaking. Living systems survive because they are inefficient by design—redundant, slow in places, ambiguous where precision would be fatal.

Optimization hates redundancy. Reality depends on it.

This is why highly optimized systems fail suddenly. They look perfect right up to the moment they are not. With no buffers left, even small disturbances exceed ATT. Resolution becomes discontinuous. Collapse feels mysterious only because warning signals were optimized away.

DELTA reframes optimization as a local tool, not a global philosophy.

Local optimization can improve performance *within bounded contexts*. Global optimization destroys coherence because it ignores interactions between subsystems. The more complex the system, the more dangerous single-metric thinking becomes.

This is the hidden cost of modern intelligence.

As our tools improve, our ability to optimize increases faster than our ability to sense secondary consequences. We become very good at pushing one curve upward while ignoring the fractures forming underneath.

The illusion of total optimization persists because early gains are real. Systems do perform better—temporarily. But performance without resilience is borrowed time. Eventually, the bill arrives as burnout, brittleness, or breakdown.

DELTA offers a quieter alternative: distributed adequacy.

Not perfect efficiency everywhere, but sufficient alignment across domains. Not maximum output, but sustainable resolution. Not the fastest path, but the one that keeps future paths open.

Abandoning total optimization is not a retreat from intelligence. It is intelligence maturing past its first success.

With this illusion dismantled, the chapter can now move toward its final synthesis: how manipulation can occur while preserving agency—how influence can shape systems without stripping them of their ability to respond.

That synthesis completes the argument for power without domination.

15.6 — Manipulation That Preserves Agency

The clearest sign that power has crossed into domination is the loss of agency in the system being influenced.

When outcomes become fixed, choices narrow, and responses are scripted, the system may still function—but it is no longer alive in any meaningful sense. DELTA treats this as a red flag. Agency is not a philosophical luxury; it is an alignment mechanism.

Agency is how systems participate in their own resolution.

Manipulation that preserves agency works *with* this mechanism instead of overriding it. It shapes conditions, gradients, and incentives without collapsing the system's internal decision loops. The goal is not obedience, but self-correcting participation.

This distinction is operational.

Domination specifies outcomes. Agency-preserving manipulation specifies boundaries and pathways.

A thermostat does not command temperature; it sets thresholds and allows fluctuation. A good teacher does not dictate conclusions; they structure inquiry. A healthy policy does not force behavior; it reshapes environments so preferred actions become naturally aligned.

In each case, delta is guided—not imposed.

Preserving agency keeps feedback alive. Systems continue to signal stress, resistance, and opportunity early. Errors remain visible. Adaptation remains possible. This is why agency-preserving interventions scale better: they distribute intelligence instead of centralizing it.

DELTA highlights a crucial asymmetry here.

Removing agency creates short-term predictability but long-term blindness. Preserving agency introduces variability, but that variability is precisely what protects coherence. Noise is not always failure; often it is early warning.

This reframing also dissolves a common fear: that respecting agency weakens power. The opposite is true.

Power that preserves agency multiplies capacity. Each participating unit becomes a local resolver of delta. The system gains redundancy, creativity, and resilience. Power that removes agency must compensate by increasing enforcement—until enforcement becomes the system's primary activity.

Agency-preserving manipulation also aligns naturally with ethics under DELTA. It respects ATT implicitly by allowing systems to resolve at their own pace. It avoids catastrophic overshoot because resistance appears early and locally, not late and globally.

This is not idealism. It is engineering at the level of living systems.

Manipulation without domination does not mean inaction. It means intervening at the level of conditions, not commands. It accepts that outcomes emerge rather than being dictated—and that emergence is not weakness, but stability.

With agency preserved, systems remain capable of learning. And learning is the only reliable antidote to future misalignment.

This brings the chapter to its final concern: what happens when alignment is ignored altogether—when early signals are dismissed, buffers removed, and agency suppressed. The answer is not moral failure.

It is structural collapse.

That cost, and how to recognize it before it becomes irreversible, is the focus of the next section.

15.7 — The Cost of Ignoring Alignment

Systems rarely fail loudly at first. They fail quietly, while everything still appears to work.

Ignoring alignment does not produce immediate chaos. It produces efficiency. Output improves. Variability drops. Resistance fades. Leaders relax. Metrics reassure. This phase is dangerous precisely because it looks like success.

DELTA explains why.

When alignment signals are ignored—when feedback is dismissed, buffers removed, and agency suppressed—delta does not vanish. It is forced into non-obvious reservoirs. Stress migrates into latency. Fatigue accumulates beneath performance. Fragility replaces flexibility.

The system becomes polite.

Polite systems do not complain. They comply. They stop signaling early discomfort and wait until resolution is unavoidable.

When collapse finally arrives, it feels sudden only because honesty was delayed.

This is why systems often break far from the point of intervention. The damage surfaces where alignment capacity was weakest, not where pressure was applied. Organizations fail at the edges. Bodies break at stress points. Societies fracture along fault lines that were long visible—but repeatedly ignored.

DELTA reframes this pattern as a simple rule: misalignment never disappears; it relocates.

Ignoring alignment also destroys learning.

When systems are punished for signaling stress, they learn to hide it. When variance is treated as failure, exploration stops. When dissent is suppressed, intelligence collapses inward. The system may continue functioning, but it becomes blind to its own degradation.

This blindness is not accidental. It is produced.

Control-heavy environments reward silence. Optimization-heavy systems reward narrow success. Over time, only information that confirms stability survives. Reality does not cooperate. It simply waits.

The cost of ignoring alignment is therefore nonlinear.

Small misalignments ignored early require massive correction later. Gentle adjustments become impossible. Resolution accelerates beyond ATT. What could have been absorbed gradually must now be released violently.

Collapse, in this sense, is not catastrophe. It is late correction.

Understanding this changes how intelligence should act. The goal is not to prevent failure entirely—that is impossible. The goal is to fail early, locally, and informatively, while alignment capacity still exists.

Respecting alignment keeps systems honest. Ignoring it makes systems brittle.

This section closes the operational argument of the chapter. Power that ignores alignment may appear decisive, but it is fundamentally insecure. Power that listens, buffers, and routes misalignment remains quiet—but durable.

With this clarity, the final synthesis becomes unavoidable.

Power is not what enforces outcomes. It is what keeps systems capable of resolving the next delta.

That realization completes this chapter—and prepares the ground for the book's final integration.

Power That Survives Its Own Use

What this chapter ultimately dismantles is a deeply comforting illusion: that power is proven by obedience.

Obedience is cheap. Coherence is expensive.

Systems obey long after they have stopped agreeing. They comply long after they have stopped adapting. By the time resistance becomes visible, alignment has already been exhausted. What follows is not rebellion—it is physics completing unfinished work.

DELTA leaves no room for romance here.

Power that survives is not the power that dominates. It is the power that keeps future resolution possible.

Manipulation is unavoidable. Every choice routes delta. Every tool amplifies reach. The question has never been whether influence should exist, but whether influence is exercised with alignment literacy or blind confidence.

Across this chapter, one pattern has repeated relentlessly:

- Control fails because it suppresses feedback.
- Optimization collapses because it removes buffers.
- Domination destroys agency and therefore intelligence.
- Ethics emerges because awareness magnifies consequence.

None of these are moral warnings. They are structural facts.

Power, properly understood, is not a peak to be reached. It is a load to be carried continuously. The stronger the system, the more carefully that load must be routed. Mastery is not force—it is pacing.

This closing does not offer comfort. It offers stability.

It replaces the fantasy of command with the discipline of stewardship. It trades short-term certainty for long-term coherence. It demands less arrogance and more accuracy.

With this chapter, the observer has been fully restored to reality: embedded, constrained, capable, and responsible. Not the creator of the world, not its ruler—but its most consequential participant.

From here, the book can move beyond correction into synthesis. The remaining chapters will integrate delta across domains—matter, life, mind, society—without granting any layer false primacy.

The journey forward is not about gaining power.

It is about learning how to use power without becoming its casualty.

True power does not impose order. It prevents misalignment from becoming fatal.

Interlude — When Explanation Stops Chasing Effects

Up to this point, the book has been doing something deliberately uncomfortable.

It has been taking familiar phenomena—motion, pressure, friction, life, mind, power—and quietly removing the explanations we are used to leaning on. Not by denying equations or observations, but by relocating causality.

For centuries, we have chased effects and called them causes.

We named forces because things moved. We named time because things changed. We named observers because experiences appeared meaningful.

Each naming solved a local problem while hiding a deeper one: what actually drives change?

DELTA has been sitting there all along—not as a number, not as a variable, not as a mystical gap—but as a condition that must resolve whenever alignment is disturbed. Once seen, it does not compete with existing laws. It organizes them.

This interlude marks the shift from correction to consolidation.

So far, we have shown where thinking runs backward: starting from force instead of imbalance, control instead of navigation, observation instead of participation. We have restored sequence—delta first, resolution next, meaning last. We have placed the observer back inside consequence, not above it.

Now something different must happen.

The final part does not introduce more examples. It introduces closure.

A law is not a claim to authority. It is a boundary of honesty—what a framework can explain, where it applies, and where it must stop. If DELTA is real, it must be able to stand still long enough to be stated plainly, without metaphor or persuasion.

This is that moment.

The next chapter does not expand the theory. It compresses it. One page. One law. Clear conditions. Clear limits. No escape clauses. No metaphysical shelter.

After that, the book removes the last illusion modern thinking clings to: that something is ever lost, that heat is waste, that silence is absence. What resolves is not energy—it is form.

And finally, we arrive at the quietest conclusion of all: a universe that needs no divine observer, no privileged frame, no mystery to function—only structure resolving misalignment, and experience arising where loops persist.

The epilogue does not persuade. It reports what changes once delta is seen everywhere.

Fear softens because chaos gains structure. Simplicity returns because excess explanation falls away. Curiosity stabilizes because mystery becomes navigable.

This final part is not a climax. It is a settling.

From here on, nothing needs to be forced.

DELTA does not promise certainty. It promises honesty that scales.

That is enough.

Part VI — A Unified Law Of Dynamism

This part completes the journey by doing what any honest framework must eventually do: *stop expanding and state itself plainly*.

This final part presents DELTA not as interpretation or philosophy, but as a scale-invariant law of dynamism—defining what drives change, how resolution proceeds, and where limits apply. The emphasis shifts from examples to structure, from narrative to compression.

Here, energy is no longer treated as something that moves or disappears, but as something that reconfigures form through alignment. Loss, waste, and dissipation are reinterpreted as terminal resolution states, not failures of conservation. Silence becomes completion, not absence.

With the law stated and consequences clarified, the last illusion falls: that reality requires mystery, divine oversight, or privileged observers to function. What remains is a universe that operates openly—through misalignment, resolution, structure, and experience—without metaphysical excess.

This part does not add more to know. It *finishes what needed to be said*.

Chapter 16 — The Delta Law (Formal Statement)

When Explanation Must Stop Expanding

Every framework reaches a point where adding more words weakens it.

Before that point, explanation feels productive. Concepts grow, examples multiply, connections widen. But beyond that point, expansion becomes avoidance—a way of postponing commitment. A law exists precisely to prevent that drift. It is not an act of authority; it is an act of restraint.

This chapter marks that restraïnt.

Up to now, DELTA has been explored through motion, fluids, friction, life, mind, power. Not to persuade, but to test whether a single causal idea can survive contact with reality across scales. It has. What remains is no longer exploration—it is statement.

A law is not a summary of observations. It is a boundary on interpretation.

Newton did not explain why gravity exists; he defined how it behaves. Thermodynamics did not describe molecules; it set limits that molecules cannot cross. In the same spirit, the Delta Law does not narrate change—it constrains it.

This chapter does not argue for DELTA. It assumes it provisionally and asks a harder question: *can it be stated without metaphor, psychology, or philosophical shelter?* If it cannot, it does not deserve to survive.

A formal statement does three things and only three things:

- It defines what must be true for a phenomenon to occur
- It defines what must never be claimed
- It defines where the law stops

Anything beyond that is commentary.

The temptation here is to be comprehensive. To show how everything fits. To demonstrate elegance. This chapter refuses that temptation. Elegance will appear if the law is correct; it cannot be engineered through language.

The Delta Law must stand on its own—brief, closed, and uncomfortable in its simplicity. It must not require belief, interpretation, or reverence. It must be usable without agreement.

That is why this chapter is intentionally austere.

No observer is privileged. No force is invoked. No scale is exempt.

If something changes, delta resolved. If nothing changes, delta did not exist.

Everything else—motion, time, heat, life, meaning—is consequence.

Once stated, the law will not be defended again in this book. It will be *used*. The next chapter will show what resolves is not energy but form. The one after will show that mystery survives only where explanation is misassigned.

This opening closes the door on narrative.

From here onward, DELTA either holds—or it does not.

There is nothing left to persuade.

16.1 — The Delta Law: Formal Statement

This section exists to do one thing only: state the law without interpretation.

Not an analogy. Not a philosophy. Not a motivation.

A law must survive being read cold.

The Delta Law (Δ)

Delta (Δ) is a condition of misalignment within a structured system. Whenever delta exists, resolution must occur. Resolution is the only physically permitted response to delta.

Formal Clarification

1. **Delta is not a quantity.**

   It is not a measurable amount of energy, force, or difference. Delta is a *state*—the presence of structural misalignment within an otherwise continuous system.

1. **Delta precedes all dynamism.**

   Motion, force, pressure, flow, transition, decay, and change do not initiate events. They are *expressions of resolution after delta already exists.*

1. **Resolution is compulsory, not optional.**

   A system cannot choose to ignore delta. If delta is present, resolution will occur immediately, gradually, discontinuously, or terminally—depending only on structural constraints and alignment capacity.

1. **Alignment governs how resolution unfolds.**

   Resolution does not seek efficiency, optimization, or purpose. It follows the available alignment pathways permitted by structure and timing.

1. **No observer is required.**

Delta resolves whether or not it is detected, measured, named, or understood. Observation may inject additional delta but does not initiate resolution.

1. **No force is fundamental.**

What is traditionally called force is a *post-resolution description* of how alignment constraints shaped the outcome.

1. **If no change occurs, no delta existed.**

Static persistence is not frozen motion. It is the absence of misalignment.

Operational Compression

- Energy exists everywhere.
- Structure defines alignment capacity.
- Delta appears as misalignment.
- Resolution follows inevitably.
- Form changes.

Nothing else is required.

Non-Claims (Explicit Exclusions)

The Delta Law does not claim:

- That delta predicts specific outcomes
- That delta replaces existing equations

- That delta introduces intention or direction
- That delta eliminates probability or uncertainty

The law constrains causality; it does not narrate behavior.

Minimal Sufficiency Test

If a phenomenon involves change, transition, flow, collapse, growth, decay, or stabilization, then:

- Delta must have existed prior to that change
- Resolution must have governed its manifestation

If either condition fails, the phenomenon is misdescribed.

This is the law in its complete form.

Nothing may be added without weakening it. Nothing may be removed without breaking it.

The next section will specify the necessary conditions under which delta can arise at all—because misalignment itself is not arbitrary.

16.2 — Necessary Conditions for Delta

Delta does not arise arbitrarily. Misalignment is not noise, accident, or fluctuation by default. It appears only when specific structural conditions are met. This section defines those conditions precisely, because without them the Delta Law would dissolve into vagueness.

Delta requires structure first.

If there is no structure—no boundary, no differentiation, no constraint—there can be no misalignment. Uniformity cannot be misaligned with itself. A perfectly homogeneous field, regardless of how much energy it contains, is dynamically silent. Energy alone does nothing.

This is the first correction modern thinking resists.

Energy is capacity. Structure is permission. Delta appears only where permission is uneven.

The necessary conditions for delta are therefore not energetic but geometric and relational.

1. Structural Asymmetry

Delta requires asymmetry in structure, not imbalance in magnitude.

Two regions may contain identical energy and still generate delta if their constraints differ. Conversely, large energy differences do nothing if the structure allows smooth continuity. Delta is born where structure prevents seamless alignment.

This is why boundaries matter.

- A membrane
- A surface
- A gradient in constraint
- A change in geometry
- A temporal mismatch

Each creates the possibility of misalignment without adding energy.

2. Constraint Differential

Structure alone is insufficient unless it limits resolution differently across regions.

Delta appears when one part of a system cannot resolve at the same rate, direction, or mode as another. This differential may be spatial (geometry), temporal (timing), or relational (coupling strength). What matters is not how much energy exists, but where resolution is restricted.

This is why uniform pressure does nothing, but pressure gradients drive flow. Why uniform temperature is inert, but gradients produce heat transfer. Why uniform charge distribution is silent, but separation generates action.

Delta is a consequence of unequal permission to resolve.

3. Continuity With Interruption

Delta cannot exist in isolated fragments. It requires continuity that is partially obstructed.

If two regions are fully disconnected, there is no misalignment—only separation. If they are fully continuous, there is no misalignment—only flow. Delta arises in the narrow regime where continuity exists *and* is constrained.

This explains why interfaces are dynamic hotspots: surfaces, films, joints, synapses, borders. They are not places where energy accumulates, but where alignment is negotiated.

4. Temporal Desynchronization

Delta may arise purely from timing.

Two subsystems resolving correctly in isolation may become misaligned when their resolution rates differ. No additional energy is required. A phase lag is sufficient. This is why oscillations drift, why feedback loops destabilize, why coordination matters more than magnitude.

Time, in DELTA, is not a backdrop. It is a resolution pacing condition.

Delta emerges when pacing mismatches exceed tolerance.

5. Non-Zero Alignment Capacity

Finally, delta requires that resolution be *possible.*

If alignment capacity is zero—if structure is rigid beyond adaptation—delta does not persist; it fractures immediately. If alignment capacity is infinite—pure fluidity—delta never forms. Delta lives only in systems that can partially adjust but not perfectly.

This is why real systems—physical, biological, cognitive—are neither rigid nor fluid. They survive in the middle, where misalignment can exist long enough to matter.

Negative Statement (What Does Not Create Delta)

- Energy presence alone
- Motion alone
- Observation alone
- Random fluctuation alone
- Difference in quantity without constraint

Without structural asymmetry and constraint differential, none of these generate delta.

Compression

Delta arises when:

- Structure exists
- Alignment is unevenly permitted
- Continuity is constrained
- Resolution pacing mismatches
- Adaptation capacity is finite

Remove any one of these, and delta collapses into silence or fragmentation.

With the conditions of delta now fixed, the next section addresses what happens after delta appears—how resolution unfolds, and why it does so in distinct regimes rather than smoothly everywhere.

That distinction is not theoretical. It governs chaos, coherence, and collapse.

16.3 — Resolution Regimes of Delta

Delta does not resolve in a single, uniform way. How misalignment unfolds depends entirely on alignment capacity, pacing, and structural constraint. Once delta appears, resolution becomes inevitable—but its *mode* is conditional.

This section defines the four resolution regimes that exhaust all physically permitted responses to delta. These regimes are not categories of systems; they are states of resolution that any system may enter as conditions change.

1. Continuous Resolution — Flow and Coherence

When alignment capacity is sufficient and pacing remains within tolerance, delta resolves smoothly.

Resolution distributes itself incrementally across structure. No accumulation occurs. No thresholds are crossed. The system appears stable, predictable, and coherent. This is the regime of flow: laminar fluids, steady motion, metabolic balance, learning curves, calm cognition.

In this regime:

- Delta exists but does not concentrate
- Feedback is immediate and local

- Structure adapts without strain

Nothing dramatic happens because nothing needs to.

Continuous resolution is not absence of delta. It is delta being handled honestly in real time.

2. Discontinuous Resolution — Jumps and Breakdown

When delta exceeds alignment capacity locally or temporally, resolution cannot proceed smoothly.

The system accumulates misalignment until a threshold is crossed. Resolution then occurs in jumps—sudden transitions, collapses, phase changes, fractures, quantum jumps, breakdowns, emotional outbursts, institutional crises.

This regime is not randomness. It is delayed resolution released all at once.

Key characteristics:

- Accumulation followed by release
- Apparent unpredictability
- Structural reconfiguration

Discontinuous resolution feels violent because pacing was ignored. But the outcome is still lawful. The system resolves the same delta it carried—just without finesse.

3. Deferred Resolution — Storage and Tension

Sometimes delta cannot resolve immediately but also cannot fracture the system.

In this case, misalignment is stored.

Stress builds. Tension accumulates. Energy appears "potential." Structures deform slightly and hold load. Examples include compressed springs, stretched membranes, social pressure, unresolved conflict, chronic stress.

Deferred resolution is not stability. It is resolution postponed under constraint.

This regime is dangerous because it is quiet. Systems appear functional while carrying hidden load. Deferred resolution always seeks completion. If release pathways are not provided, it will migrate toward discontinuity.

4. Terminal Resolution — Dispersion and Silence

When alignment capacity is exhausted entirely, resolution reaches its final state.

Delta disperses into incoherent modes. Structure loses the ability to carry further misalignment. Energy spreads uniformly. No gradients remain. No further resolution is possible.

This is terminal resolution.

Examples include:

- Heat dissipation
- Mechanical failure
- Death of biological systems
- Collapse of meaning
- Silence after exhaustion

Terminal resolution is not destruction. It is completion.

Nothing remains unresolved, and therefore nothing remains dynamic.

What These Regimes Clarify

- Chaos is not a cause; it is a regime

- Order is not a goal; it is a regime
- Stability is not permanence; it is pacing
- Collapse is not anomaly; it is delayed honesty

Systems move between regimes continuously. A single system may host multiple regimes simultaneously across scales.

Compression

Delta resolves as:

- Flow, when alignment capacity is sufficient
- Jumps, when thresholds are crossed
- Tension, when resolution is deferred
- Silence, when resolution is complete

There is no fifth mode.

With the regimes defined, the law now requires its boundaries. A law that claims to explain everything explains nothing. The next section defines where the Delta Law stops, what it cannot do, and why those limits are essential rather than embarrassing.

That discipline keeps the law honest.

16.4 — Limits of the Delta Law

A law earns credibility not by what it claims, but by what it refuses to claim.

The Delta Law is intentionally bounded. Its strength lies in its restraint. This section defines those boundaries clearly, because without limits, delta would collapse into metaphor—and metaphor is where rigor goes to die.

1. Delta Does Not Predict Specific Outcomes

Delta explains why change must occur, not *what exact form it will take.*

Once misalignment exists, resolution is inevitable—but the pathway depends on structure, constraints, timing, and local contingencies. Two systems with identical delta may resolve differently because their alignment landscapes differ.

Delta governs necessity, not detail.

Prediction remains the domain of equations, simulations, and statistics. Delta only guarantees that *something must resolve.* It does not choose how.

2. Delta Does Not Eliminate Probability or Uncertainty

Probability is not ignorance of delta. It is indeterminacy within resolution pathways.

Even when delta is known, systems with multiple viable alignment routes may resolve through any of them. Noise, microstructure, and timing determine which path is taken. Delta does not remove uncertainty; it frames where uncertainty lives.

Randomness, therefore, is conditional—not fundamental, but not illusory either.

3. Delta Does Not Replace Mathematics

The Delta Law is pre-mathematical, not anti-mathematical.

Mathematics describes how resolution unfolds *once variables are defined.* Delta defines why variables must change at all. These roles are complementary, not competitive.

Equations remain indispensable for:

- Quantification
- Prediction
- Engineering
- Precision

Delta simply prevents mistaking equations for causes.

4. Delta Does Not Create Structure

Delta operates within structure; it does not generate it from nothing.

Structure determines alignment capacity. Without structure, delta cannot exist. The law therefore does not explain the origin of geometry, constants, or fundamental architectures. It explains what happens once those exist and misalign.

Origin questions remain separate.

5. Delta Does Not Grant Control or Authority

Understanding delta does not confer power over systems.

It increases sensitivity to consequence, not exemption from it. Alignment limits remain binding. ATT remains finite. Misuse of insight accelerates failure rather than preventing it.

Delta offers honesty, not dominance.

6. Delta Is Not a Theory of Meaning or Purpose

Meaning arises *after* resolution, not before it.

The Delta Law does not assign goals, values, or intent to nature. Any meaning perceived is the observer's internal compression of resolved patterns—not a directive embedded in reality.

Purpose is experiential, not causal.

Why These Limits Matter

Without these boundaries:

- Delta would become unfalsifiable
- Explanation would slide into narrative
- Physics would blur into philosophy

With them, the law remains clean, usable, and non-mystical.

Compression

The Delta Law:

- Explains why change is unavoidable
- Explains how resolution must occur
- Refuses to explain everything else

This refusal is not weakness. It is discipline.

With the limits fixed, the law can now be placed properly among existing frameworks—showing what it can and cannot do—before the book moves on to its final reinterpretation of energy, loss, and silence.

That placement is the work of the next section.

16.5 — What Delta Can and Cannot Do

With the law stated and its limits fixed, what remains is practical clarity. This section separates capability from temptation—what the Delta Law legitimately explains, and what it must never be stretched to cover.

This separation is not academic. It is what keeps DELTA usable rather than mythic.

What Delta Can Do

1. Initiate Manifestation, Delta explains why nothing happens until misalignment appears. Energy may exist everywhere, but without delta there is no trigger, no direction, no necessity for change. Manifestation begins at the moment alignment is disturbed.

This alone corrects centuries of force-first thinking.

2. Drive Motion Without Force Language, Motion does not require an abstract push. It follows resolution pathways opened by misalignment. What we later call force is a *description of constraint response*, not the driver itself.

Delta allows motion to be understood as consequence, not command.

3. Explain Transitions Across Scales, The same misalignment–resolution logic governs quantum jumps, fluid flow, biological metabolism, cognition, and social change. Constants differ. Structures differ. But the causal grammar remains identical.

This is not unification by reduction. It is unification by condition.

4. Clarify Stability and Collapse, Delta explains why systems remain stable only when resolution keeps pace—and why collapse always follows deferred misalignment. Stability is not permanence; it is continuous honesty. Collapse is not mystery; it is delayed accounting.

5. Integrate Physics, Life, and Mind Without Hierarchy, No domain is privileged. Matter, life, and mind are not stacked realities but different resolution regimes of the same dynamism. Meaning does not drive physics. Physics does not trivialize meaning. Each appears where structure allows.

Delta dissolves false hierarchy without flattening difference.

What Delta Cannot Do

1. Delta Cannot Create Structure, Misalignment requires structure to misalign against. Delta operates *within* geometry, boundaries, and constraints. It does not explain why those exist, only how they behave once uneven.

Origin remains outside its scope.

2. Delta Cannot Predict Specific Outcomes, Knowing delta exists does not tell you exactly what will happen—only that something must. Outcomes depend on structure, timing, and local conditions. Delta guarantees resolution, not form.

Prediction remains a secondary tool.

3. Delta Cannot Eliminate Uncertainty, Uncertainty arises from multiple viable resolution pathways. Delta frames this uncertainty; it does not abolish it. Randomness is conditional, not mystical—but it is real.

4. Delta Cannot Override Alignment Limits, No amount of insight allows systems to resolve faster than their ATT. Knowledge does not grant exemption. Attempting to bypass alignment limits accelerates failure.

Delta offers warning, not permission.

5. Delta Cannot Replace Existing Laws, Conservation laws, thermodynamics, mechanics, and statistical models remain valid within their domains. Delta does not compete with them. It precedes them.

It explains why they must operate, not how they calculate.

6. Delta Cannot Assign Meaning or Purpose, Meaning appears after resolution, inside observers. Delta does not encode intent. Nature does not aim. Purpose is an experiential compression, not a causal directive.

Compression

Delta can:

- Explain *why* change must occur
- Explain *when* stability breaks
- Explain *why* domains align structurally

Delta cannot:

- Explain *everything*
- Predict *exact outcomes*
- Grant *control* or *purpose*

This asymmetry is essential.

A framework that claims omnipotence collapses under its own ambition. Delta survives because it remains structurally modest.

With this clarity in place, the final task of this chapter remains: to show why delta is not limited by scale—why it operates identically from quantum jumps to civilizational shifts.

That scale-invariance is addressed next.

16.6 — Scale Invariance of Delta

If the Delta Law were confined to a single scale, it would be another domain theory. Its strength lies in the opposite claim: delta does not care about scale.

What changes with scale are structures, constants, and descriptive languages—not the causal condition itself. Misalignment resolves whether the system is microscopic or planetary, inert or living, physical or cognitive. The grammar of change remains intact.

This is what scale invariance means here: the same condition produces change everywhere, though it manifests differently.

Delta Is Blind to Size

Delta does not measure meters, joules, seconds, or neurons. It responds only to misalignment within permitted structure.

A quantum jump occurs because resolution cannot proceed continuously. A fluid flows because resolution can. A biological system lives because it sustains internal delta loops. A society transforms because misalignment exceeds institutional ATT.

Different phenomena. Same cause.

The differences we assign—classical vs quantum, physical vs biological, objective vs subjective—are descriptive partitions, not causal ones. Delta crosses them without resistance.

Constants Change, Conditions Do Not

At different scales:

- Geometry changes
- Constraints change
- Resolution bandwidth changes
- Time perception changes

But the condition that *something must resolve when misalignment exists* does not.

This is why no special clause is required for:

- Quantum discontinuity
- Turbulence in fluids
- Sudden insight in cognition
- Social tipping points

Each is simply a regime shift caused by delta–ATT mismatch at that scale.

The constants governing the system differ. The necessity of resolution does not.

Why Scale Breaks Equations but Not Delta

Equations are scale-sensitive because they depend on measurable quantities. As scale changes, dominant variables shift, approximations fail, and new terms appear. This is expected.

Delta does not break because it is pre-quantitative.

It does not calculate outcomes. It constrains possibility. It says only this: *if misalignment exists and resolution is possible, change must occur.*

That statement remains true whether resolution takes femtoseconds or centuries.

No Privileged Scale, No Privileged Domain

Scale invariance removes the need for hierarchy.

- Quantum is not more fundamental than classical
- Physics is not more real than biology
- Biology is not more basic than mind

Each is a resolution theater operating under the same law with different structural rules.

There is no deepest layer where delta suddenly stops working. There is no emergent layer where delta suddenly begins. It is present wherever structure allows misalignment to persist long enough to matter.

Compression

Delta is scale-invariant because:

- Misalignment does not depend on size
- Resolution does not require measurement
- Structure, not magnitude, governs outcome

What changes with scale is how resolution appears, not *why it occurs*.

This completes the formal statement of the Delta Law.

Nothing essential remains unstated.

The next chapter moves forward—not by adding another principle, but by removing a persistent misunderstanding: the idea that what resolves is energy itself.

What resolves is form.

Energy remains conserved. Form does not.

Understanding that difference dissolves the final confusion around loss, heat, and silence.

16.7 — Delta and Existing Laws (Clarification, Not Conflict)

The Delta Law does not compete with existing laws of physics. It sits beneath them.

Conflict arises only when descriptive tools are mistaken for causal engines. Mechanics, thermodynamics, and field theories are exquisitely successful at *describing how outcomes unfold once variables are in motion*. The Delta Law addresses a different question: why motion must occur at all when misalignment exists.

This distinction is not cosmetic. It restores order.

Delta and Conservation Laws

Conservation laws describe what remains invariant through change. Delta explains why change begins.

Energy conservation is untouched. Momentum conservation is untouched. Charge conservation is untouched. Delta does not add or remove quantities; it introduces necessity. When misalignment appears, resolution must redistribute conserved quantities along permitted pathways.

Conservation governs accounting. Delta governs initiation.

There is no violation because they answer different questions.

Delta and Thermodynamics

Thermodynamics explains directionality through entropy gradients. Delta explains why gradients appear and must resolve.

Entropy increase is not the cause of change; it is a *terminal description* of how resolution disperses when alignment capacity is exhausted. Delta does not contradict the Second Law—it explains why irreversible processes arise once misalignment cannot be resolved coherently.

Heat is not wasted energy. It is terminal alignment dispersion.

Delta and Mechanics

Classical mechanics begins with force. Delta ends the need to treat force as primitive.

Forces are constraint responses observed after misalignment is resolved through structure. Newton's laws remain valid because they describe consistent outcomes of constrained resolution. Delta simply removes the metaphysical burden placed on force as a driver.

Mechanics predicts trajectories. Delta explains why trajectories exist.

Delta and Field Theories

Field theories describe how quantities vary across space and time. Delta explains why variation becomes active.

A uniform field does nothing. A gradient does. Delta identifies the condition under which gradients must resolve. Field equations then specify *how* that resolution propagates.

No field equation is replaced. Its causal grounding is clarified.

Delta and Quantum Theory

Quantum mechanics formalizes discontinuity. Delta explains why continuity fails when alignment bandwidth is insufficient.

Quantum jumps are not mysterious decisions. They are resolution events in delta-starved systems where continuous pathways do not exist. Probability remains because multiple resolution routes remain viable.

Delta does not remove quantum indeterminacy. It removes the need for observer-driven metaphysics.

Delta as Pre-Law Condition

Delta is not a higher law. It is a pre-law condition—a causal prerequisite that makes laws applicable.

Laws operate *after* delta appears. Delta operates *before* laws are invoked.

This positioning prevents overreach. Delta does not explain everything; it explains why explanation becomes necessary.

Compression

- Existing laws describe outcomes
- Delta constrains initiation
- No law is violated
- No law is replaced

Delta restores causal sequence without rewriting mathematics.

With this placement complete, the framework is closed.

What follows in the next chapter is not another principle, but a correction of language: what actually resolves is not energy, but form. Understanding that distinction dissolves the final confusion around loss, heat, and silence—and prepares the ground for the book's quiet conclusion.

A Law That Knows Where to Stop

The Delta Law is complete precisely because it does not try to be complete.

It does not explain origins. It does not predict outcomes. It does not replace equations, observers, or meaning.

It does something narrower—and therefore stronger.

It restores causal order.

Change does not begin with force. It does not begin with energy. It does not begin with observation.

Change begins when misalignment exists within structure and resolution becomes unavoidable.

Everything that follows—motion, flow, heat, life, cognition, collapse—is consequence.

By placing delta before laws without competing with them, the framework avoids the two failures that plague grand theories: overreach and mystification. It neither claims dominion over all phenomena nor retreats into metaphor when pressed.

Delta is not a new actor in nature. It is the condition that makes actors move.

Once this is seen, many confusions dissolve quietly:

- Forces stop pretending to be causes.
- Entropy stops pretending to be intention.
- Observation stops pretending to be creation.
- Collapse stops pretending to be anomaly.

What remains is a universe that behaves honestly—without drama, without privilege, without exception.

Energy is conserved. Structure constrains. Misalignment resolves.

Nothing more is required.

With this law placed, the book can now release its final burden: the idea that something is ever truly lost when systems change. The next chapter completes the picture by showing that what resolves is form, not energy—and that silence is not absence, but completion.

After that, there will be nothing left to correct.

Only a different way of seeing.

Energy exists everywhere. Nothing happens until delta resolves.

Chapter 17 — What Resolves Is Form, Not Energy

The Misplaced Fear of Loss

Much of our discomfort with change comes from a single, deeply rooted assumption: that something is lost when systems transform.

We speak of "energy loss," "wasted heat," "dissipation," "decay," as if reality were constantly misplacing parts of itself. This language is so familiar that we rarely question it. Yet it hides a contradiction. Physics insists that energy is conserved, while our intuition keeps insisting that something vanishes.

The contradiction is not in nature. It is in where we locate loss.

What disappears during change is not energy. It is form—the specific arrangement that once carried alignment in a particular way. When that arrangement can no longer sustain misalignment, it resolves. To the observer attached to that form, resolution feels like loss. To nature, it is simply completion.

Energy never goes missing. It spreads, redistributes, equalizes. What ends is the structure that once shaped its pathways. We confuse the disappearance of pattern with the disappearance of substance, and that confusion carries emotional weight far beyond physics.

This is why endings feel heavier than beginnings. A beginning creates form. An ending dissolves it. Even when nothing is destroyed, the *recognizable organization* is gone. Our fear is not of annihilation, but of unrecognizable continuity.

DELTA reframes this gently but firmly.

Resolution does not consume energy. It finishes a configuration.

Heat is not failure. Silence is not emptiness. Stillness is not death. They are states where alignment has reached its limit, where no gradients remain to demand further change. Nothing is broken. Nothing is missing. Nothing is trying to happen anymore.

This chapter exists to relocate loss from physics back into perception—where it belongs.

Once that relocation is made, many ideas soften at once: dissipation becomes honest, entropy loses its menace, and silence regains dignity. What remains is a universe that never leaks, never forgets, and never panics—only one that continuously rearranges form until nothing remains unresolved.

From here, the fear of loss no longer drives explanation.

Only clarity does.

17.1 — Energy Does Not Resolve, Form Does

Energy does not end. It does not exhaust itself. It does not "go away."

What ends—what resolves—is form.

This distinction is subtle, and because it is subtle, it has been missed repeatedly. Physics has always known that energy is conserved, yet our explanations behave as if energy were being consumed, depleted, or lost. This contradiction persists because we keep assigning the act of resolution to the wrong entity.

Resolution is not something energy does. Resolution is something structure undergoes.

Energy is capacity—the ability to participate in change. It is indifferent to configuration. Form, on the other hand, is a *specific alignment* of that capacity within constraints. When misalignment exists, it is the form that carries tension, stores gradients, and channels resolution. When that form can no longer sustain misalignment, it changes or dissolves.

Energy simply follows.

Consider any familiar process: a stretched spring relaxing, a hot object cooling, a moving body coming to rest, a bubble bursting, a living organism dying. In none of these cases is energy destroyed. What disappears is the organized pathway through which energy was being expressed.

The spring loses its shape. The temperature gradient disappears. The velocity field dissipates. The membrane ruptures. The metabolic loop breaks.

Energy remains present in every case—redistributed, dispersed, equalized—but the *form that once held it coherently is gone*. We mistakenly describe this as energy resolving because we focus on measurable quantities. What actually resolved was the alignment configuration that made those quantities meaningful.

DELTA makes this unavoidable.

Misalignment exists only relative to form. Energy without form cannot be misaligned. It cannot carry tension. It cannot demand resolution. A perfectly uniform energy field is dynamically silent—not because energy is absent, but because nothing is structured enough to resist alignment.

This is why change always appears where structure exists.

Mountains erode because they have shape. Bubbles burst because they have membranes. Societies collapse because they have institutions. Identities dissolve because they have narratives.

Energy does not care about any of this. Form does.

Once form is recognized as the true site of resolution, a long-standing confusion evaporates. We stop asking where energy "went" and start asking which structure could no longer carry misalignment. That question is always answerable.

Resolution is not disappearance. It is reconfiguration or completion of form.

This insight prepares the ground for the next correction. If form resolves while energy remains, then conservation laws have never been violated—only misinterpreted. Understanding that reinterpretation is the task of the next section.

17.2 — Reinterpreting Energy Conservation

Energy conservation has never been wrong. What has been wrong is what we expected conservation to explain.

Classical and modern physics insist—correctly—that energy is neither created nor destroyed. Yet this insistence coexists with a persistent discomfort: systems slow down, heat up, decay, and settle into states that feel *poorer* than before. We call this loss because we are watching form degrade, not because energy has vanished.

DELTA resolves this tension by separating accounting from causality.

Conservation laws answer the question: *How much capacity remains in the system?* DELTA answers the question: *Why did the system have to change its configuration?*

These are orthogonal questions.

When a moving object comes to rest due to friction, no energy is lost. Kinetic energy is redistributed into microscopic motion—heat, sound, deformation. What is lost is the coherent form of motion: the organized velocity field that once carried alignment across the object and its surroundings.

We mistake the disappearance of coherence for disappearance of substance.

Energy conservation tells us *nothing disappeared*. DELTA tells us *what finished resolving*.

This is why conservation alone feels incomplete as an explanation. It accounts for totals, not for why ordered patterns dissolve into dispersed ones. Entropy fills part of that gap descriptively, but DELTA places the cause earlier: misalignment within form must resolve, and resolution often ends in dispersion when structure can no longer sustain gradients.

Seen this way, conservation laws are bookkeepers, not narrators.

They track invariants through change. They do not explain why change begins, why it proceeds irreversibly, or why particular forms cannot be recovered without new alignment work. DELTA supplies that missing narrative without violating a single conservation principle.

This reinterpretation also clarifies a subtle point: conservation does not imply recoverability.

Energy can be conserved while form is irretrievably altered. Once a gradient is dispersed and alignment pathways collapse, reversing the process requires new delta introduced deliberately—new structure, new boundaries, new constraints. Conservation guarantees possibility in principle, not feasibility in practice.

That is why heat engines need temperature differences. Why memory requires structure. Why life requires continuous alignment work.

Energy conservation remains intact. What changes is our expectation that conservation should preserve usefulness, order, or meaning.

DELTA makes the boundary explicit:

- Conservation preserves quantity
- Resolution transforms quality

Once this is understood, the idea of "loss" can be relocated properly. Nothing is ever lost in the physical sense—but something very specific *does* end.

That ending is the completion of a form's ability to carry misalignment.

The next section makes this relocation explicit by addressing the most emotionally loaded claim of all: the belief that something is permanently lost when systems dissipate.

17.3 — Why Nothing Is Ever Lost

The idea that something is lost during change survives not because it is correct, but because it is psychologically efficient.

When a system transforms and can no longer do what it once did, we compress that experience into a single word: loss. The word feels accurate because it matches our inability to *use* the system as before. But inability to use is not disappearance. It is misrecognition.

Nothing is ever lost in physical reality. What is lost is access to a former alignment.

DELTA makes this distinction unavoidable.

When a system resolves delta, it does not destroy energy. It redistributes it into configurations that no longer support the previous form. From the perspective of that form, the world has ended. From the perspective of the larger system, nothing has vanished—only coherence has dispersed.

This is why loss feels asymmetric.

Gaining form is slow and deliberate. Losing form feels sudden and final.

Creation requires alignment work. Dissolution requires only that alignment work stop. Once gradients flatten and structures relax, energy spreads into modes that are harder to reconcentrate. Not impossible—just costly. The cost is not energetic annihilation, but organizational reconstruction.

This is what irreversibility actually means.

Irreversibility is not about energy disappearing. It is about the collapse of pathways that once allowed energy to move coherently.

Entropy describes this statistically. DELTA explains it structurally. When form resolves, the specific constraints that made gradients usable dissolve. Energy remains present but is now expressed across too many microstates to be recollected without fresh intervention.

That intervention is always external.

To rebuild form, new boundaries must be introduced. New gradients must be created. New misalignment must be deliberately injected. Conservation alone cannot do this. Resolution finished its work. What follows requires new structure.

This is why nothing is ever truly lost, yet almost everything that matters must be rebuilt.

Memory, once dispersed, is not destroyed—but it is no longer accessible. Heat, once equalized, is not gone—but it is no longer directional. A life, once ended, has not annihilated energy—but the form that sustained internal resolution loops has completed.

DELTA removes the metaphysical sting from this realization.

There is no cosmic waste bin. There is no erasure. There is only completed alignment.

Loss exists only from the perspective of a form that can no longer continue resolving delta. Nature does not mourn this. It does not celebrate it either. It simply proceeds, because nothing remains unresolved.

Understanding this reframes endings completely.

They are not failures. They are not punishments. They are not mysteries.

They are points at which a form has said everything it could say.

The next section addresses the most misunderstood endpoint of all—heat—and shows why it is not waste, not loss, but the most honest expression of terminal resolution.

17.4 — Heat as Terminal Alignment Dispersion

Heat has long been treated as the graveyard of useful energy.

We call it waste. We design systems to minimize it. We speak of "loss to heat" as if something valuable slipped through our fingers. This language persists even though physics insists—correctly—that nothing has vanished.

DELTA resolves this contradiction by naming heat for what it actually is: terminal alignment dispersion.

Heat is not failed energy. It is energy released from form.

When a structure can no longer sustain misalignment—when gradients collapse, constraints relax, and pathways dissolve—energy does not disappear. It spreads into the most statistically uniform configuration available. That uniformity is what we call heat.

Heat appears not because energy is excessive, but because structure is insufficient.

Consider friction. Two surfaces in relative motion carry organized alignment: velocity fields, directional gradients, constrained contact pathways. As misalignment resolves, those pathways break down into microscopic vibrations. Energy disperses across countless microstates. The organized form of motion ends. Heat appears.

Nothing is wasted. Everything is finished.

This is why heat is irreversible in practice. Once energy is evenly dispersed, there are no gradients left to drive coherent motion. To extract work again, one must introduce new structure—a temperature difference, a boundary, a cycle. Heat engines do not consume heat; they exploit new misalignment introduced elsewhere.

DELTA makes this precise.

Heat is the state reached when:

- Alignment capacity is exhausted
- Resolution pathways are saturated
- No directional gradients remain

It is not disorder in the moral sense. It is honest completion.

This reframing also clarifies why heat is universal. Every resolving process, if allowed to run fully, ends here. Motion slows. Gradients flatten. Form dissolves. Energy remains—but without preference, without direction, without urgency.

Calling heat "waste" reflects our attachment to form, not a failure of nature.

From the system's perspective, heat is not a problem.

From the observer's perspective, it marks the end of usefulness for a specific configuration.

This distinction matters beyond physics. Burnout, fatigue, boredom, saturation—all mirror the same structure. When alignment capacity is exceeded and no new pathways exist, resolution disperses. Energy remains, but coherence is gone.

Heat is therefore not an enemy of order. It is the boundary of what order can sustain.

Once this boundary is respected, the final state becomes intelligible rather than ominous. When all gradients are gone, when nothing presses to change, what remains is not chaos.

It is silence.

Understanding silence as full resolution—not absence—is the final correction this chapter must make.

17.5 — Silence as Full Resolution

Silence is usually misunderstood as emptiness.

We associate it with absence—of motion, of sound, of meaning. In doing so, we project our dependence on activity onto reality itself. DELTA requires a different reading. Silence is not what remains when something fails to happen. It is what remains when nothing needs to happen.

Silence is full resolution.

When all misalignment has resolved—when gradients are gone, constraints relaxed, and no differential demands continuation—dynamism completes itself. Energy remains present, but it is evenly expressed. There is no direction left to follow, no tension left to release, no pathway privileged over another.

This state is not inertness by lack. It is stability by completion.

In physics, this appears as thermal equilibrium. In mechanics, as rest. In acoustics, as stillness. In cognition, as calm clarity. In systems, as saturation. Different languages, same condition: no unresolved delta remains.

Silence does not oppose motion. It succeeds it.

This reframing removes the metaphysical weight we often attach to stillness. Silence is not death in the cosmic sense. It is the point at which a particular form has finished expressing all the alignment it could sustain. What follows—if anything—requires new delta, introduced by new structure or external interaction.

This is why silence is stable but not terminal in an absolute sense.

Silence persists until misalignment reappears. When structure changes, delta returns. Resolution begins again.

There is no tragedy in this cycle. There is no narrative arc imposed by nature. Silence is simply the honest end of a process.

DELTA also clarifies why silence feels unsettling to observers.

Human systems evolved to detect change because unresolved delta once meant threat or opportunity. Silence carries no such signal. It offers no instruction. To a mind conditioned for action, that feels like emptiness. In reality, it is freedom from demand.

Nothing presses. Nothing pulls. Nothing must be done.

Seen this way, silence is not the opposite of life. It is what life touches briefly when its internal deltas are aligned. Calm is not the absence of energy; it is energy without misalignment.

This understanding completes the correction begun in this chapter.

- Energy is conserved.

- Form resolves.
- Heat disperses alignment.
- Silence completes the process.

Nothing is lost. Nothing escapes. Nothing hides behind mystery.

With this, the final emotional residue of "loss" leaves the framework. Change no longer implies disappearance. Endings no longer imply negation. They imply completion.

The next chapter can now say its final sentence honestly: a reality that needs no divine observer, no metaphysical escape, and no unresolved fear—only structure, resolution, and experience where loops persist.

After that, the book will not explain more.

It will simply let you see.

17.6 — The Emotional Residue of Form Loss

Even after understanding that nothing is physically lost, something still aches.

That ache is not irrational. It is form memory.

Humans do not grieve energy. We grieve configurations—faces, routines, identities, roles, narratives, bodies, meanings. Our emotional systems evolved to attach to stable forms because forms once guaranteed continuity. When a form dissolves, the emotional system registers it as threat, even when no physical loss has occurred.

DELTA does not dismiss this reaction. It locates it correctly.

Emotion is an internal resolution process responding to the collapse of a familiar alignment. When a form that organized expectation disappears, the mind experiences a sudden surplus of unresolved delta. There is no longer a pathway for prediction, control, or continuation. That surplus manifests as grief, fear, nostalgia, or resistance.

This is not weakness. It is physics expressed internally.

The mistake occurs when emotional residue is projected outward—when the mind concludes that because a form has ended, reality itself has diminished. That projection turns completion into tragedy and resolution into injustice.

Form loss feels final because reconstruction requires work.

To rebuild alignment, new structures must form. New gradients must appear. New meaning loops must stabilize. Until that happens, the system remains in a partially unresolved state. Emotion is simply the signal that internal alignment has not yet caught up with external completion.

This is why time matters in grief.

Not because reality needs time—but because internal resolution bandwidth is finite. Emotional systems have their own ATT. They cannot absorb abrupt form loss instantly. Resolution must proceed gradually, or it becomes discontinuous and overwhelming.

DELTA reframes emotional pain not as resistance to truth, but as delayed alignment.

Understanding that nothing is lost does not eliminate emotion. It removes metaphysical panic from it. The pain is no longer about annihilation. It is about transition. The work is not to recover what vanished, but to permit new form to emerge.

This reframing also explains why some losses never fully resolve.

Certain forms are foundational to identity. When they dissolve, the system that depended on them must reorganize at a deep level. That reorganization may take years—or may never fully stabilize in the same way again. This is not failure. It is structural consequence.

DELTA offers no consolation here—only honesty.

Form loss hurts because form mattered. It mattered because it carried alignment. It ended because it finished resolving what it could.

Nothing was stolen. Nothing escaped. But something meaningful completed its role.

Recognizing this does not erase pain. It prevents pain from becoming illusion—from turning completion into cosmic injustice.

With the emotional residue now properly placed, the chapter can close its final loop: from misalignment, to resolution, to silence—without fear, without mystery, and without denial.

That completion is the work of the final section.

17.7 — Completing the Picture of Dynamism

With form, heat, silence, and emotion now placed correctly, dynamism can finally be seen whole—without gaps, without ghosts, without unfinished explanations.

Dynamism is not endless motion. It is finite resolution occurring wherever misalignment exists.

Every dynamic process follows the same arc:

1. Structure permits alignment
2. Misalignment (delta) appears
3. Resolution proceeds within capacity
4. Form adapts, transforms, or dissolves
5. Energy disperses uniformly
6. Silence remains until new delta appears

Nothing interrupts this arc. Nothing escapes it. Nothing needs to be added to complete it.

What confused us for centuries was our tendency to break this arc into pieces and then ask why they did not explain one another. We separated motion from structure, energy from form, entropy from emotion, endings from meaning—and then wondered why reality felt fragmented.

DELTA closes that fragmentation.

Change is not driven by force. Order is not defeated by entropy. Endings are not failures.

They are resolution completing itself honestly.

This section matters because it dissolves the last refuge of mystery: the belief that something fundamental must be missing when explanation feels quiet. We are conditioned to expect drama at the bottom of things. When none appears, we assume depth has been lost.

In truth, depth has been compressed.

Dynamism does not need perpetual tension to remain real. It needs only the possibility of misalignment. When misalignment ends, dynamism rests. That rest is not absence—it is availability. The system is free to host new structure, new gradients, new stories.

This understanding changes how decay, death, and endings are interpreted.

Decay is form releasing burden. Death is a loop completing. Endings are not negations; they are closures that make space.

Nothing mystical remains to be explained away. Nothing tragic remains to be justified. Reality does not require consolation because it never loses anything.

What remains unresolved is not physics—but our attachment to unfinished form.

When that attachment loosens, what appears is not emptiness, but clarity. The universe no longer feels hostile or indifferent. It feels *precise*. It does exactly what it must, no more and no less.

With this, the chapter is complete.

There is no longer a question of where energy goes, why heat appears, or what silence means. Those questions dissolve once form is placed at the center of resolution.

Only one step remains in the book: to remove mystery entirely—not by denying wonder, but by showing that wonder does not require obscurity.

That final step belongs to the next chapter.

From Resolution to Simplicity

Once form is recognized as the true site of resolution, the universe stops leaking.

Nothing is misplaced. Nothing vanishes. Nothing escapes into mystery.

What ends is only a way of being arranged.

Energy remains continuous throughout all change, indifferent to the shapes it temporarily inhabits. Form appears where alignment can be sustained and dissolves when it cannot. Heat is not waste but completion. Silence is not emptiness but rest.

This understanding does not strip reality of meaning—it returns meaning to the correct scale. Meaning belongs to observers because observers compress resolved patterns into experience. Nature does not need meaning to function. It needs only structure and misalignment.

With this correction, fear quietly exits the framework.

Fear depended on the idea of loss. Loss depended on the idea of disappearance. Disappearance depended on misunderstanding what resolves.

Once that chain breaks, endings soften. Change loses its menace. Stillness regains dignity.

What remains is a world that operates openly—without hidden realms, without metaphysical escape hatches, without privileged witnesses. Nothing waits to be seen. Nothing hides behind collapse.

Resolution happens. Form completes. Energy remains.

The next chapter removes the last residual temptation: the urge to reintroduce mystery when explanation feels calm. It shows that reality needs no divine observer, no secret frame, and no metaphysical surplus to remain profound.

After that, the book will not argue further.

It will simply invite you to see the world in delta.

Energy remains. Form resolves. Silence completes.

Chapter 18 — Reality Without Mystery

When Mystery Was Carrying What Clarity Should

For a long time, mystery did important work for us.

It filled gaps where explanation failed. It softened fear where understanding ended. It gave meaning to silence when stillness felt unbearable. Mystery was not foolish—it was functional. But over time, it began carrying weight that did not belong to it.

We started using mystery not to honor the unknown, but to protect unresolved confusion.

Whenever explanation reached a quiet end, we added something beyond it: a hidden observer, a deeper layer, a final secret. We mistook the absence of friction for the absence of depth. Clarity felt thin, so we padded it with transcendence.

DELTA changes that posture.

When misalignment resolves fully, explanation naturally stops. Nothing is missing. Nothing is waiting to be uncovered. Silence is not concealment; it is completion. Mystery only appears when we expect motion where none is required.

This chapter does not attack wonder. It separates wonder from obscurity.

Wonder arises when structure reveals itself cleanly—when complexity collapses into intelligible simplicity. Obscurity arises when causality is misassigned and gaps are treated as sacred. The two feel similar at first glance, but they pull in opposite directions.

Reality does not become shallow when it becomes readable. It becomes honest.

For centuries, we carried the idea that something must stand behind the world—watching it, choosing outcomes, holding it together. That idea persisted not because it was true, but because we had not yet learned how to describe change without intention, order without command, meaning without authorship.

Now we can.

This chapter does not remove mystery by force. It lets mystery retire gracefully. Nothing is being taken away. Nothing essential is being denied. What disappears is only the habit of adding invisible actors when explanation finishes its work.

What remains is a reality that does not need guardians, permissions, or metaphysical scaffolding to function.

Just misalignment. Just resolution. Just structure. Just experience.

Once this is accepted, the world does not feel smaller.

It feels complete.

18.1 — No Divine Observer

The idea of a divine observer did not arise from arrogance. It arose from misplaced causality.

When early thinkers noticed that outcomes changed upon interaction—especially in measurement—they inferred presence behind process. When order appeared persistent, they inferred intention sustaining it. When silence followed resolution, they inferred a watcher holding things still.

Each inference made sense locally. None survived structurally.

DELTA removes the need for a divine observer without diminishing reverence for reality. It does so by restoring sequence.

Nature does not wait to be seen. It resolves because misalignment exists.

Observation enters after that fact.

A divine observer was introduced to answer a question that no longer needs asking: *what keeps reality going when nothing appears to be happening?* The answer is now simple—nothing needs to keep it going. Silence is not suspended action. It is completed resolution.

This reframing dissolves a long-standing confusion between awareness and stability.

Stability does not require supervision. Order does not require intention. Continuity does not require attention.

Mountains remain standing without witnesses. Atoms bind without approval. Fields interact without consent. The universe has never paused to check if it was being observed—and no experiment has ever required it to.

The divine observer concept also fails under scale.

If observation creates reality, whose observation mattered before life emerged? If awareness sustains existence, what sustained awareness itself? Each attempt to preserve the observer introduces infinite regress—observers watching observers, endlessly deferred.

DELTA cuts through this cleanly.

Resolution does not need to be seen. Seeing is simply another resolution event.

Observers are products of stabilized internal delta loops, not overseers of the external world. They arise when systems become complex enough to reference their own alignment states. This capacity grants experience—not authority.

This does not strip the universe of meaning. It relocates meaning where it belongs.

Meaning does not govern reality. Reality permits meaning.

Once this is accepted, the universe stops feeling like a performance and starts feeling like a process. Nothing is trying to impress. Nothing is hiding outcomes. Nothing is watching to approve or disapprove.

Reality does not need validation to proceed.

By removing the divine observer, we do not remove depth—we remove unnecessary burden. We stop asking *who is watching?* and start asking *what is resolving?*

That shift is not reduction. It is precision.

With the illusion of a cosmic watcher released, the next correction becomes possible: removing the idea that any perspective—human or otherwise—occupies a privileged frame.

That clarity is the work of the next section.

18.2 — No Privileged Frame

The second illusion that keeps mystery artificially alive is the belief that some viewpoint owns reality.

Sometimes that viewpoint is divine. Sometimes it is human. Sometimes it is mathematical, cultural, or historical.

All versions share the same mistake: confusing a useful frame with a privileged one.

DELTA removes this confusion by grounding reality in resolution, not perspective.

A frame is a tool. It is never an authority.

Every observation, measurement, and description occurs from within constraints—location, scale, timing, structure, language. These constraints shape *how* resolution is experienced, not *what* resolves. Change does not ask which frame is correct before it occurs. It resolves wherever misalignment exists, according to local structure.

This is why multiple frames can disagree without reality breaking.

A particle looks like a wave in one setup and a particle in another—not because reality is undecided, but because different alignment injections sample different resolution pathways. A river looks calm from afar and turbulent up close—not because it changed its nature, but because scale alters what misalignment becomes visible.

Relativity formalized this mathematically. DELTA completes it causally.

No frame is privileged because no frame escapes alignment limits. Each frame reveals some gradients while hiding others. Each trades precision for blindness in a different direction. What persists across frames is not description, but necessity: delta resolves.

This also dissolves the temptation to elevate experience into truth.

Human perception feels central because it is internally coherent. That coherence gives the illusion of completeness. But internal coherence is not global authority. It is simply successful local resolution. Other systems—cells, ecosystems, stars—resolve just as honestly without sharing our viewpoint.

Removing privileged frames does not lead to relativism.

Relativism claims all descriptions are equally true. DELTA claims all descriptions are equally limited.

Reality is not negotiable, but access to it always is.

This distinction matters because privileged frames breed metaphysics. When one viewpoint is elevated above others, everything outside it must be explained away, transcended, or declared illusory. Mystery becomes a defensive move.

DELTA requires no such defense.

Reality is not hidden behind perspectives. It is sampled through them.

Once this is accepted, explanation can stop cleanly. There is no need to hunt for a final vantage point where everything is revealed at once. There is no "view from nowhere" waiting behind physics.

There are only structured systems resolving misalignment, each from where it stands.

With divine oversight removed and privileged frames dissolved, only one temptation remains: the urge to escape explanation entirely by invoking something beyond it.

That urge—and why it is unnecessary—is addressed next.

18.3 — No Metaphysical Escape

When explanation reaches its natural end, many feel an urge to step outside it.

Not because something is wrong—but because completion feels unfamiliar.

For centuries, metaphysics functioned as an escape hatch. Whenever causal chains stopped, whenever equations fell silent, whenever resolution reached stillness, we assumed something *beyond* must take over. A deeper layer. A hidden cause. A transcendent principle.

DELTA removes the need for that escape by showing that explanation is allowed to finish.

Silence at the end of explanation is not failure. It is correct termination.

Metaphysical escapes appear precisely where resolution has completed but expectation has not adjusted. We confuse the absence of further causation with the absence of reality. Instead of accepting that nothing more is required, we invent a realm where answers are postponed indefinitely.

This habit turns limits into mysteries.

When physics cannot say more, we say "beyond physics." When causality closes, we say "beyond causality." When resolution ends, we say "something unseen must continue."

But nothing continues because nothing needs to.

DELTA makes this explicit: causality operates only while misalignment exists. Once alignment is complete, there is no hidden engine running quietly in the background. There is only stability.

Stability does not require explanation any more than rest requires force.

Metaphysical escape also persists because humans mistake meaning-seeking for cause-seeking.

Meaning belongs to experience. Cause belongs to structure.

When cause ends, meaning may continue internally—but that continuation does not imply an external mechanism sustaining it. We project our inner narrative outward and call the projection "deeper reality."

This projection creates unnecessary anxiety.

If something must always lie beyond explanation, then understanding is never safe. If mystery must be preserved, clarity becomes suspect. If silence must be filled, peace feels incomplete.

DELTA rejects this posture gently.

There is no "beyond" waiting to be entered. There is only this, fully resolved or actively resolving.

This does not diminish awe. It refines it.

A universe that functions without hidden scaffolding is not impoverished—it is precise. A reality that does not need metaphysical padding is not shallow—it is stable. Mystery, when overused, becomes noise added after alignment has already occurred.

DELTA does not abolish metaphysics by argument. It renders it unnecessary by sufficiency.

Once misalignment, resolution, structure, and experience account for everything that happens, nothing remains to be outsourced to another realm. No invisible guarantees are required. No transcendent explanations are owed.

Explanation stops because it should stop.

With no divine observer, no privileged frame, and no metaphysical escape left standing, reality finally rests on its own feet—unsupported, unprotected, and complete.

What remains is not emptiness, but plainness.

The next section shows why this plainness is not reduction, but the most faithful description reality can have.

18.4 — Just Resolution

Once the last metaphysical escape is closed, what remains may feel almost too simple.

Not simplistic—but unadorned.

Reality does not proceed by intention, narrative, or purpose. It does not aim, choose, or hesitate. It does not seek outcomes or avoid them. What it does—consistently, relentlessly, and without exception—is resolve misalignment whenever it exists.

That is all.

Resolution is not a principle layered on top of nature. It is what nature does when structure cannot remain uneven.

This reframing removes a deep, inherited habit: the urge to assign agency where none is required. We have been trained to expect decisions behind change—forces pushing, fields commanding, laws enforcing, observers collapsing possibilities. DELTA shows that none of these are necessary.

Change happens because not changing is no longer permitted by structure.

Resolution has no preference. It does not optimize, moralize, or pursue elegance. It follows available pathways defined by geometry, constraints, and timing. When pathways are smooth, resolution flows. When pathways are blocked, resolution accumulates. When accumulation exceeds capacity, resolution jumps. When capacity ends, resolution disperses.

No drama. No story. No exception.

This does not reduce reality. It frees it from narrative burden.

A process that requires intention must be justified. A process that follows necessity does not.

Resolution is sufficient cause. Nothing else needs to be added behind it—not will, not intelligence, not hidden guidance. The moment misalignment exists, resolution is already underway. The moment misalignment ends, resolution rests.

This explains why nature appears both creative and indifferent.

Creation emerges where structure allows sustained resolution. Indifference appears where resolution has completed.

Neither requires explanation beyond alignment dynamics.

Understanding reality as "just resolution" also stabilizes thought. It removes the anxiety that something must always be happening behind the scenes. When nothing changes, nothing is secretly working. Silence is honest. Stillness is earned.

There is profound relief in this.

It means the universe is not fragile. It does not need supervision. It does not depend on constant activity to remain real.

Resolution happens when required. Otherwise, reality rests.

Once this is accepted, the remaining elements fall into place naturally. Resolution needs structure to operate. Structure creates the stage on which experience becomes possible.

That brings us to the next clarification: just structure—and why it matters without ruling anything.

18.5 — Just Structure

Resolution does not occur in a vacuum. It requires structure—not as an overseer, but as a constraint that makes pathways real.

Structure is often mistaken for rigidity, scaffolding, or imposed order. In reality, structure is simply the geometry of permission. It defines what can happen, how it can happen, and how quickly it can happen—without deciding *that* it should happen.

Structure does not command resolution. It shapes its routes.

Every boundary, interface, lattice, membrane, surface, gradient, and constraint is a structural statement. It says: *resolution may pass this way, but not that way*. Change follows these statements automatically. No instruction is needed.

This is why the same energy behaves differently in different contexts.

A river erodes rock slowly because the rock's structure limits alignment pathways. Air flows easily because its structure offers little resistance. Soap films form spheres because the boundary geometry equalizes gradients. Living cells sustain activity because membranes, enzymes, and cycles keep resolution looping internally.

Structure is not passive. But it is not active either.

It does not push. It permits.

DELTA places structure at the center without turning it into a hidden agent. Structure does not store intention or carry purpose. It simply constrains misalignment until resolution finds the available route. When structure changes, routes change. When routes disappear, resolution disperses.

This also explains why structure feels "real" in a way abstractions do not.

Energy is invisible until structured. Resolution is intangible until constrained. Experience does not exist without boundaries.

Structure is where possibility becomes outcome.

Once this is understood, many false debates dissolve. Order versus chaos becomes a question of structural bandwidth. Stability versus change becomes a question of structural endurance. Control versus freedom becomes a question of how rigidly pathways are enforced.

No metaphysics is required to explain why structure matters. It matters because without it, nothing can resolve coherently.

And yet, structure never rules the process. It never dictates meaning. It never elevates itself into authority. It simply stands there—quietly determining what is possible.

This is why structure precedes meaning.

Meaning arises after resolution has passed through structure and been experienced internally. Nature does not encode meaning into form. Meaning is compressed *from* form by systems capable of internal coherence.

That distinction prepares the final clarification of this chapter: just experience—why experience matters deeply without governing reality.

That is the work of the next section.

18.6 — Just Experience

Experience is real—but it is not foundational.

This distinction matters more than it appears.

For centuries, philosophy and science oscillated between two extremes: either experience creates reality, or experience is an illusion riding on something more "objective." Both positions miss the same point. Experience is neither ruler nor ghost. It is a consequence of sustained internal resolution.

Experience happens inside systems that can align themselves.

When a system becomes complex enough to reference its own state—when internal deltas loop rather than collapse immediately—experience emerges. Sensation, awareness, memory, anticipation, emotion, meaning: all are internal alignment phenomena, not external authorities.

DELTA places experience exactly where it belongs.

Experience does not make reality happen. Reality makes experience possible.

This corrects a deep psychological bias. Because experience is immediate, vivid, and continuous, it feels primary. The world seems to arrive *through* experience, so we mistake the interface for the source. But an interface does not generate what it displays.

Experience is resolution seen from within.

Pain is internal delta resisting resolution. Pleasure is internal alignment flowing smoothly. Attention is resolution bandwidth being allocated. Meaning is compressed coherence after alignment stabilizes.

None of these govern the external world. They govern how the external world is navigated internally.

This is why experience can be intense without being authoritative.

A dream feels real while it lasts. Fear feels absolute while unresolved. Belief feels final while coherent.

Yet none of these alter the underlying resolution dynamics outside the system experiencing them.

Recognizing this does not trivialize experience. It liberates it.

Experience no longer needs to carry cosmic responsibility. It does not have to justify reality, sustain existence, or explain causation. It only needs to do what it does best: guide local navigation.

This also dissolves the false war between objectivity and subjectivity.

Objectivity describes resolution independent of experience. Subjectivity describes resolution internalized as experience.

They are not rivals. They are different projections of the same dynamism.

Once this is accepted, the observer can finally be put back in place—not elevated, not dismissed, but integrated. The observer becomes what it always was: a participant system, resolving its own internal deltas while embedded in larger ones.

Experience matters because it is where life is lived. It does not rule because ruling was never its function.

With experience correctly placed—real but not sovereign—the framework is complete. Nothing remains inflated, diminished, or misplaced.

The final section of this chapter closes the loop: explaining why, once everything is put back in its proper role, explanation itself can stop without loss.

That completion is not an ending.

It is readability restored.

18.7 — The End of Explanatory Inflation

When every misplaced role is returned to its proper place, something unexpected happens: explanation becomes quiet.

Not incomplete. Not evasive. Just finished.

Explanatory inflation is the habit of adding layers when none are required—of mistaking clarity for shallowness and simplicity for omission. We were trained to believe that if an explanation feels light, it must be missing something heavy. So we added intention, observers, deeper realms, hidden agencies, and final purposes—each meant to thicken reality where understanding had already settled.

DELTA reverses this habit.

It shows that when misalignment resolves, nothing remains to explain. The urge to continue explaining is not driven by reality, but by discomfort with closure. We confuse the absence of further cause with the absence of meaning.

But meaning is not a property of explanation. It is a property of experience.

Once resolution, structure, and experience are placed correctly, explanation can stop without leaving residue. There is no dangling "why" demanding metaphysical rescue. There is no unfinished business hiding behind silence.

This is not reductionism. It is completionism.

A completed explanation does not answer every possible question. It answers the necessary ones and refuses the rest. It respects boundaries. It does not promise transcendence where none is required.

This is why the framework now feels stable.

Nothing needs protection. Nothing needs supplementation. Nothing needs mystery to remain significant.

Reality is not fragile; it does not collapse when explanation ends. It stands on its own because it always did.

This section matters because it releases the reader from an inherited pressure: the pressure to keep digging when the ground is already solid. Curiosity remains—but it becomes grounded curiosity, not anxious excavation.

You can still explore. You can still wonder. But you no longer need to inflate.

When explanation stops honestly, what replaces it is not ignorance—it is readability. The world becomes something you can live inside without metaphysical tension. Change no longer implies threat. Stillness no longer implies void.

With explanatory inflation ended, the chapter has nothing more to add.

What remains is a synthesis—not a new claim, but a quiet confirmation that reality does not need to be protected by mystery to be profound.

That synthesis closes the chapter.

A World That Does Not Need Mystery to Survive

What remains after everything unnecessary is removed is not emptiness. It is stability.

Reality does not depend on hidden witnesses, privileged perspectives, or metaphysical scaffolding to function. It does not need intention to move, purpose to organize, or mystery to feel deep. It requires only what it has always used: misalignment, structure, and resolution.

Nothing more was ever doing the work.

The divine observer was a placeholder for unresolved causality. The privileged frame was a substitute for limited access. Metaphysical escape was a refuge from honest completion.

Each was understandable. None was necessary.

Once they fall away, reality becomes readable again—not in fragments, not through special access, but directly. Change occurs because it must. Stillness remains because it can. Experience matters because systems capable of alignment live inside it.

There is no final secret withheld at the bottom of things. There is no deeper layer waiting to be unlocked once the right belief is held. The universe does not conceal itself—it simply does not perform.

This does not eliminate wonder. It removes anxiety from wonder.

Wonder no longer depends on obscurity. Depth no longer depends on confusion. Meaning no longer depends on authority.

What you are left with is a world that does not demand faith, surrender, or metaphysical courage. It only asks for attention to what resolves and why.

Reality without mystery is not a world explained away. It is a world explained enough.

And that is sufficient.

With this chapter complete, the book has nothing left to prove. What remains is not argument, but perspective—the effect of seeing dynamics through delta.

The epilogue will not add concepts. It will describe what quietly changes once you stop looking for what was never missing.

Reality is not mysterious because it hides. It is intelligible because nothing needs to be hidden.

Epilogue — Seeing The World In Delta

Something subtle changes once delta is seen.

Not dramatically. Not suddenly. Quietly—like realizing you were holding your breath for years and have finally exhaled.

You stop asking *why* things happen in a moral or cosmic sense, and start noticing how misalignment resolves. Events no longer feel personal in the way they once did. They feel contextual. Situations stop appearing as rewards or punishments and begin to reveal themselves as consequences of structure meeting delta.

This does not make life colder. It makes it calmer.

Fear reduces—not because uncertainty disappears, but because mystery no longer threatens. When you understand that nothing vanishes, nothing conspires, and nothing waits behind the curtain, anxiety loses its metaphysical fuel. Change is no longer interpreted as loss of meaning, only as completion of form.

You begin to notice where delta is too large—where chaos is inevitable no matter how much force is applied. You also notice where delta is too small—where stagnation, noise, or pointless fluctuation masquerades as progress. Gradually, almost unintentionally, you start respecting the right delta.

Patience stops feeling like virtue. It starts feeling physical.

You sense when systems need time—not as advice, not as ethics, but as bandwidth. You stop rushing processes that cannot resolve faster without breaking coherence. You stop forcing clarity where alignment has not yet matured.

Curiosity changes too.

It becomes steadier. Less hungry. More precise. You no longer chase explanations to exhaustion, because you trust that explanation is allowed to end. When understanding reaches silence, you do not panic. You recognize silence as full resolution, not ignorance.

This has consequences for how you see yourself.

You no longer imagine yourself as an exception—neither divine nor insignificant. You are an organized system, resolving internal deltas while embedded in larger ones. Your thoughts, emotions, ambitions, and doubts are not enemies to defeat, but signals of alignment in motion.

Even suffering changes shape.

It is still painful—but it is no longer mysterious. You see it as delayed resolution, constrained structure, or misaligned expectation. That does not eliminate pain, but it prevents it from becoming existentially overwhelming. Suffering stops asking metaphysical questions it was never meant to answer.

The world itself begins to feel strangely honest.

Mountains stand without meaning. Rivers flow without intention. People act within the limits of their alignment capacity. Systems fail when delta exceeds what structure can hold.

Nothing feels staged. Nothing feels judged. Nothing feels secretly hostile.

And yet—everything feels alive.

Not because it is animated by purpose, but because wherever misalignment exists, resolution is already underway. Life is not special because it is chosen. It is special because it sustains internal delta loops long enough to experience them.

Seeing the world in delta does not turn you away from science, philosophy, or wonder. It makes them lighter. You stop demanding more than they can give. You stop turning explanations into beliefs and beliefs into burdens.

You begin to live inside reality rather than negotiating with it.

That is all this book was ever meant to do.

Not to replace existing theories. Not to overthrow meaning. Not to sell mystery under a new name.

Only to restore a simple, quiet truth:

Nothing manifests because energy exists. Manifestation begins when delta starts resolving.

Once you see that, you do not unsee it.

You simply move through the world a little more aligned— and let the rest resolve on its own.

Closing Note

I did not write this book to introduce a new belief.

I wrote it because, over years of teaching, observing, arguing with equations, and quietly watching life unfold, I kept running into the same discomfort: *our explanations worked, but our intuition didn't settle.* Something always felt slightly off—especially when theory met reality in the most ordinary moments.

- A soap bubble.
- Traffic.
- Fatigue.
- Learning.
- Conflict.
- Silence.

These are not exotic problems. Yet they kept exposing the same crack: we were starting our explanations too late—from force, motion, reaction—while missing what actually *triggers* dynamism.

Delta.

Once I saw it clearly, I could not unsee it. And once I framed everything around delta—misalignment with the capacity to resolve—many long-standing confusions simply stopped demanding attention. Not because they were answered dramatically, but because they were no longer the right questions.

This book is not an attack on physics, philosophy, or spirituality. It is a request to place each of them where they belong.

Physics does not need intention to work. Philosophy does not need mystery to feel deep. Life does not need force to move forward.

Most of what troubles us—personally and collectively—comes from delta blindness: pushing where alignment is needed, accelerating where bandwidth is limited, moralizing what is structurally inevitable.

If this book has done anything worthwhile, I hope it has done just this:

- Replaced force with sensitivity
- Replaced urgency with bandwidth
- Replaced mystery with clarity
- Replaced fear with orientation

You do not need to agree with every framing here. You do not need to adopt the language permanently. What matters is whether, after reading, the world feels more readable.

If you now pause more often before forcing outcomes... If you sense when patience is not virtue but physics... If you recognize silence as completion rather than absence...

Then the book has already served its purpose.

Delta does not ask to be believed. It asks to be noticed.

Everything else will resolve on its own.

Appendices

"Once you see the world in delta, you stop forcing outcomes and start respecting coherence."

Appendix A — Glossary of Core Terms

This glossary is not a dictionary of symbols. It is a *calibration tool*—meant to keep words aligned with reality as used in this book. Each term is defined operationally, not poetically.

Delta (Δ)

Delta is misalignment with the capacity to resolve.

Delta is not a numerical difference, not a force, and not a quantity to be conserved. It is a condition—the state in which a system is uneven enough to demand change, yet structured enough to allow resolution.

Delta exists wherever:

- Gradients exist
- Constraints prevent uniformity
- Alignment is incomplete

Delta precedes force, motion, time, and causality. Without delta, nothing happens—even if energy exists.

Resolution

Resolution is the process by which delta reduces through available pathways.

Resolution is not a decision, intention, or optimization. It is what occurs automatically when misalignment exists within structure.

Resolution can appear as:

- Motion (when pathways are continuous)
- Jumps (when pathways are discontinuous)
- Dissipation (when pathways saturate)
- Silence (when delta is fully resolved)

Resolution always moves in one direction: toward alignment. When resolution ends, change ends—honestly.

Alignment Threshold Time (ATT)

ATT is the maximum rate at which a system can resolve delta without losing coherence.

ATT is system-specific. It depends on:

- Structure
- Scale
- Geometry
- Internal feedback capacity

When delta exceeds ATT:

- Systems destabilize
- Chaos, turbulence, or breakdown occurs

When delta is far below ATT:

- Noise dominates
- Fluctuations replace flow

ATT explains why patience is physical, not moral—and why forcing speed often causes failure.

Coherence

Coherence is sustained resolution without structural breakdown.

A system is coherent when:

- Delta is present
- Resolution proceeds within ATT
- Structure remains intact

Coherence is not stillness. It is controlled dynamism.

Living systems, stable flows, learning processes, and meaningful conversations all depend on coherence. Loss of coherence does not mean failure—it means delta exceeded what structure could carry.

Ripple Tension

Ripple tension is alignment pressure within a boundary, not a pulling force.

Ripple tension replaces the classical idea of surface tension in this framework. It describes how alignment propagates through molecular or field structures to minimize unresolved delta.

Key characteristics:

- It is not force-based
- It arises from collective alignment, not attraction
- It reduces surface area because spherical boundaries equalize gradients

Soap films, membranes, interfaces, and boundaries stabilize due to ripple tension—not because molecules "pull," but because alignment seeks equipotential structure.

Form

Form is a temporary configuration that sustains alignment.

. Form is not energy. Form is what energy *passes through* during resolution.

Forms appear where structure allows sustained alignment and dissolve when it does not. Nothing "owns" form—not matter, not energy, not observers.

Heat

Heat is terminal alignment dispersion.

Heat is not wasted energy. It is energy released from structured pathways once alignment capacity is exhausted.

Heat marks completion, not failure.

Silence

Silence is full resolution with no remaining delta.

Silence is not emptiness or absence. It is stability without demand.

Silence persists until new delta appears.

Observer

An observer is a system capable of internal resolution loops.

Observers do not create reality. They experience resolution internally and assign meaning post-alignment.

Experience is real. Authority is not.

Meaning

Meaning is compressed coherence after resolution.

Meaning does not drive reality. It is extracted *from* resolved patterns by observers.

Summary Anchor

- Energy exists.
- Delta invites change.
- Resolution produces form.
- Structure shapes pathways.
- Coherence sustains flow.
- Heat completes alignment.
- Silence rests honestly.
- Experience remembers.

Appendix B — Comparative Text

Force-Based vs Delta-Based Physics

This appendix does not attempt to replace classical physics. Instead, it repositions causality. Force-based physics describes outcomes accurately but often misidentifies what initiates change. Delta-based physics retains the mathematics but shifts the starting point of explanation.

Core Comparison

In force-based physics, the primary driver is the action of force on objects. In delta-based physics, the driver is misalignment (delta), which demands resolution. Motion in the classical framing arises because a force is applied; in the delta framing, motion begins because misalignment within a structure cannot persist.

Energy in the force-based view is fuel for motion, while in the delta-based view it is the capacity that participates in resolution. Equilibrium in classical physics is treated as a special or ideal state, whereas in delta-based physics equilibrium is the default whenever delta equals zero.

Change happens in the classical model because force is applied; in the delta model, it happens because structures cannot remain misaligned. The direction of change is determined by net force in the classical view, but by available resolution pathways in the delta view. Time is treated as an independent parameter in force-based physics, while in delta-based physics it is a measure of resolution progression.

Instability in the classical framing is failure of balance; in the delta framing it is delta exceeding ATT (available resolution bandwidth). Dissipation is seen as energy loss in the classical model, but as alignment diffusion in the delta model. The end state in force-based physics is zero force or rest, while in delta-based physics it is silence—full resolution.

Key Conceptual Shifts

1. **From "Pushing" to "Resolving"** Force language assumes an external agent pushes systems into motion. Delta language recognizes that motion begins internally when gradients exist that cannot persist. Nothing needs to be pushed once misalignment exists.
2. **From Objects to Conditions** Classical models focus on objects as primary entities. Delta-based framing treats conditions—gradients, constraints, interfaces—as primary. Objects are stabilized forms, not initiators.
3. **From Balance to Bandwidth** Classical stability asks whether forces balance. Delta-based stability asks whether resolution bandwidth (ATT) is sufficient. Systems fail not when balance is imperfect, but when delta exceeds what structure can absorb per unit time.
4. **From Loss to Completion** Traditional models speak of losses (frictional loss, head loss, energy loss). Delta-based framing treats these as completion of structured resolution, not disappearance. Nothing is lost; form finishes.
5. **From Absolute Time to Process Time** Force-based physics treats time as a backdrop. Delta-based physics treats time as what resolution feels like when observed. Different systems have different ATT, leading to different effective timescales.

Why Mathematics Still Works

This reframing does not invalidate equations:

- Newton's laws still predict motion.
- Navier–Stokes still models flows.
- Laplace's law still relates pressure and curvature.
- Bernoulli still conserves along streamlines.

What changes is interpretation:

- Equations describe how resolution proceeds, not why it begins.
- Constants encode structural limits, not universal intent.
- Boundary conditions become causal, not secondary.

Mathematics was never wrong—it was context-incomplete.

Practical Consequences

- **Engineering:** Focus shifts from forcing performance to managing delta and ATT.
- **Fluid Mechanics:** Pressure gradients replace "push" intuition.
- **Friction & Wear:** Treated as alignment diffusion, not enemy forces.
- **Learning Systems:** Speed limits explained structurally, not motivationally.
- **Human Systems:** Conflict seen as delta mismanagement, not moral failure.

One-Line Summary

Force describes motion. Delta explains why motion becomes unavoidable.

Appendix C — Everyday Examples

Seeing Delta at Work

This appendix translates DELTA out of theory and into lived reality. Each example follows the same pattern:

Structure → Delta → Resolution → Outcome

Once this pattern is seen, it becomes difficult to unsee.

1. Soap Bubble — Why Gentle Blowing Works

- **Structure:** Soap solution forms a flexible molecular boundary (film) with ripple tension.
- **Delta:** A *very small pressure gradient* between inside air and surrounding air.
- **Resolution:** If delta is introduced slowly and stays within the film's ATT, alignment propagates evenly across the boundary. The film reorganizes into a spherical equipotential surface.
- **Outcome:** A stable bubble forms and moves dynamically to maintain equilibrium with external air currents. When delta exceeds film capacity, the boundary ruptures and resolution disperses as droplets and sound.

Key Insight: Bubbles do not form because of force. They form because delta is introduced within alignment bandwidth.

2. Traffic Flow — Why Speeding Creates Jams

- **Structure:** Roads, lanes, signaling systems, vehicle spacing.
- **Delta:** Differences in speed, braking, and lane-change behavior.
- **Resolution:** When delta remains moderate, traffic flows coherently. When delta spikes (sudden braking, aggressive overtaking), resolution cannot propagate smoothly and collapses into stop–go waves.
- **Outcome:** Traffic jams appear *without accidents*. Energy (cars, engines) still exists—but flow coherence is lost.

Key Insight: Congestion is not caused by "too many cars." It is caused by delta exceeding traffic ATT.

3. Learning — Why Forcing Speed Backfires

- **Structure:** Cognitive capacity, prior knowledge, attention bandwidth.
- **Delta:** Gap between new information and existing understanding.
- **Resolution:** When delta is moderate, learning integrates smoothly. When delta is too large, confusion and anxiety appear. When delta is too small, boredom and noise dominate.
- **Outcome:** Effective learning occurs only at right delta.

Key Insight: Learning failure is not lack of intelligence. It is delta mismanagement.

4. Conflict — Why Arguments Escalate

- **Structure:** Relationships, trust, communication channels.
- **Delta:** Mismatch in expectations, values, or emotional states.
- **Resolution:** If delta is acknowledged and addressed gradually, dialogue stabilizes. If delta is denied or amplified, resolution jumps into aggression or withdrawal.
- **Outcome:** Conflicts "explode" not due to disagreement, but because resolution bandwidth is exceeded.

Key Insight: Conflict is not moral failure. It is unresolved delta in relational structure.

5. Fatigue & Burnout — Why Rest Is Not Laziness

- **Structure:** Body, nervous system, metabolic cycles.
- **Delta:** Accumulated mismatch between demand and recovery.
- **Resolution:** When delta builds slowly, the system compensates. When delta exceeds physiological ATT, resolution collapses into exhaustion.
- **Outcome:** Burnout appears suddenly—but it was structurally inevitable.

Key Insight: Fatigue is not weakness. It is alignment debt coming due.

6. Innovation — Why Pressure Kills Creativity

- **Structure:** Mental space, exploratory freedom, safety.
- **Delta:** Demand for novelty under constraints.
- **Resolution:** Moderate delta stimulates insight. Excessive

delta collapses creativity into imitation or paralysis.
- **Outcome:** Innovation thrives only where delta is challenging but survivable.

7. Leadership & Power — Why Control Fails

- **Structure:** Institutions, incentives, norms.
- **Delta:** Mismatch between imposed rules and lived reality.
- **Resolution:** Heavy-handed control increases delta faster than structure can absorb. The system appears stable—until breakdown occurs.
- **Outcome:** Collapse feels sudden but was structurally guaranteed.

Key Insight: Power fails not from opposition, but from delta blindness.

8. Personal Growth — Why Change Feels Hard

- **Structure:** Identity, habits, belief systems.
- **Delta:** Mismatch between current self and desired self.
- **Resolution:** Gradual change integrates. Forced transformation fractures identity.
- **Outcome:** Lasting growth occurs only through coherent resolution, not force.

Pattern Summary

Across all examples:

- Too much delta → chaos
- Too little delta → stagnation
- Right delta → coherence

This pattern is universal because it is structural, not psychological.

Life does not respond to force. It responds to delta within structure.

Appendix D — Roadmap

How DELTA Connects to UED, Universal Skill, and Human Systems

This appendix situates DELTA inside a larger, coherent ecosystem of ideas. It answers a practical question the reader may not yet know how to ask:

"If delta explains dynamism, where does it live in the bigger picture?"

The answer is: everywhere—but at different layers.

1. DELTA as the Entry Point (Local Causality Layer)

DELTA is the most accessible layer of the framework.

It explains *why something starts happening* in any system—physical, biological, psychological, or social—without requiring advanced mathematics or metaphysics.

At this layer:

- Delta explains initiation
- Resolution explains direction
- ATT explains limits
- Coherence explains stability

DELTA is deliberately minimal. It does not attempt to explain the universe. It explains dynamism where you are standing.

Think of DELTA as the lens you put on before analysis begins.

2. UED (Universal Energy Dynamics) — The Structural Field Layer

UED sits beneath DELTA.

If DELTA explains *when and why change begins*, UED explains what the universe is made of such that delta can exist at all.

Within UED:

- Energy is structured, not chaotic
- Vacuum is not empty but organized
- Fields are not forces but alignment media
- Matter is condensed resolution

DELTA operates *inside* UED as a local expression of global structure.

Relationship:

- UED defines the terrain
- DELTA explains movement on that terrain

You do not need UED to use DELTA. But UED explains *why DELTA works everywhere.*

3. Universal Skill — The Human Alignment Layer

Universal Skill is the human-facing translation of DELTA.

Where DELTA explains system behavior, Universal Skill explains how humans interact with systems intelligently.

Universal Skill is the capacity to:

- Sense delta accurately
- Respect ATT (in self and others)
- Avoid force-based intervention
- Enable coherent resolution

At this layer:

- Poor decisions = delta blindness
- Wisdom = delta literacy
- Stress = unresolved internal delta
- Mastery = alignment timing

Universal Skill is not habit, discipline, or motivation. It is structural sensitivity.

Relationship:

- DELTA explains reality
- Universal Skill explains competent participation in reality

4. Human Systems — The Collective Dynamics Layer

Human systems (education, organizations, markets, politics, families) are multi-agent delta systems.

They fail not because people are bad, but because:

- Delta is misdiagnosed
- ATT is ignored
- Force replaces alignment

DELTA reframes systemic issues cleanly:

- Burnout → delta accumulation
- Conflict → misaligned expectations
- Collapse → delta exceeding structural bandwidth
- Reform → structural realignment, not pressure

At this layer:

- Ethics becomes delta responsibility

- Leadership becomes misalignment management
- Policy becomes ATT engineering

Relationship:

- DELTA explains system stress
- Universal Skill enables individuals to navigate it
- Structural design determines outcomes

5. Why This Book Comes First

This book intentionally starts before UED, philosophy, psychology, or social theory.

Because:

- Force-first thinking blocks deeper insight
- Metaphysical shortcuts create confusion
- Human systems copy physical misconceptions

DELTA clears the ground.
Once delta is understood:

- UED becomes intuitive, not abstract
- Universal Skill becomes obvious, not moral
- Human systems become readable, not frustrating

This is why DELTA is not a theory of everything.
It is a key.

6. Suggested Reading Path (Optional)

For readers entering the ecosystem:

1. DELTA — Learn how change actually begins
2. UED works — Understand the structured universe

beneath

3. Universal Skill books — Apply delta intelligence to life
4. Human systems / geopolitics / psychology — See scale-invariant consequences

Each layer assumes the previous—not as belief, but as *calibration*.

DELTA explains why change begins. UED explains why change is possible. Universal Skill explains how to act wisely within change. Human systems reveal what happens when we ignore all three.

Quote Bank

Selected Lines from DELTA

1. *"Energy is capacity. Delta is invitation."*
2. *"Nothing manifests because energy exists—manifestation begins only when delta starts resolving."*
3. *"Force describes motion; delta explains why motion becomes unavoidable."*
4. *"Resolution is not a decision. It is what structure does when misalignment can no longer persist."*
5. *"Silence is not absence. It is full resolution."*
6. *"Nothing is ever lost in nature—only the form that once carried alignment completes."*
7. *"Heat is not wasted energy; it is energy released from form."*
8. *"Chaos is not disorder—it is delta beyond a system's capacity to resolve."*
9. *"Patience is not a moral virtue. It is a physical bandwidth."*
10. *"Equilibrium is not special. It is simply what remains when delta is gone."*
11. *"Reality does not wait to be observed. Observation only experiences what has already resolved."*
12. *"A system does not fail because it is weak, but because delta exceeded what its structure could hold."*
13. *"Friction is not resistance to motion—it is alignment diffusion."*
14. *"Learning fails not from lack of intelligence, but from delta mismanagement."*

15. *"Structure does not command change; it merely permits certain resolutions and forbids others."*
16. *"Meaning is not a cause. It is compressed coherence after alignment."*
17. *"Nothing needs to keep the universe running—silence is not suspended action, it is completion."*
18. *"Loss exists only from the perspective of a form that can no longer continue."*
19. *"The observer is not divine, not irrelevant—just an organized participant in resolution."*
20. *"When explanation ends cleanly, mystery is no longer required."*
21. *"Control fails because force ignores delta."*
22. *"Stability is not the absence of change; it is change resolving within bandwidth."*
23. *"Reality is not mysterious because it hides—it is intelligible because nothing needs to be hidden."*
24. *"Delta blindness turns alignment problems into moral dramas."*
25. *"Once you see the world in delta, you stop forcing outcomes and start respecting coherence."*

Acknowledgements

This book did not emerge from isolation, even though much of the thinking behind it happened quietly.

I owe my first acknowledgement to **every student** who asked an uncomfortable question, looked unconvinced after a correct derivation, or sensed that an equation worked while intuition lagged behind. Those moments mattered. They revealed where explanation was complete on paper but incomplete in understanding. DELTA was shaped in that gap.

I am grateful to the **classroom itself**—not as a place, but as a testing ground. Teaching the same concepts repeatedly forces honesty. Anything artificial collapses quickly. Anything real survives repetition. DELTA survived.

I acknowledge the **everyday moments** that carried more insight than formal settings ever could: watching bubbles form, observing fatigue, noticing traffic patterns, sensing silence after resolution. Reality was the most patient teacher in this work.

I thank those who engaged in **serious disagreement** with me. Not for agreement, but for pressure. Pressure exposed weak framing and forced clarity. DELTA sharpened where it was challenged, not where it was praised.

I am deeply grateful to my **family**, who lived alongside this thinking long before it became a manuscript. Especially my children, whose ordinary questions often cut through layers of abstraction effortlessly—and reminded me that reality does not require complexity to be true.

This book also stands on the shoulders of **physics, philosophy, and human inquiry**, even where it reframes them. I do not reject what came before. I reposition it. Any clarity here belongs as much to centuries of thinkers as it does to my own stubborn refusal to ignore misalignment.

Finally, I acknowledge **silence**.

Many of the insights in this book did not arrive while thinking harder, but while stopping at the right moment—when explanation had finished and did not need rescue. DELTA owes its existence to learning when to stop pushing.

If this book feels calm rather than convincing, that is intentional.

Clarity does not argue loudly. It settles.

— **Sandeep J. Chavan**

About the Author

Sandeep J. Chavan is an independent researcher, educator, and prolific author whose work spans physics, philosophy, psychology, and human systems. Trained as an engineer and known locally as a rigorous physics teacher, he is recognized for challenging force-centric explanations and replacing them with structurally grounded, reality-first frameworks.

Chavan is the originator of **Universal Energy Dynamics (UED)** and related alignment-based models that reinterpret motion, stability, intelligence, and human behavior without metaphysical shortcuts. His writing emphasizes clarity over mystique, consequences over beliefs, and structure over speculation—making complex ideas readable without diluting rigor.

With more than seventy published books and multiple research reports, his work forms a coherent ecosystem addressing topics from fundamental physics and consciousness to education, leadership, and societal resilience. Across domains, a single throughline remains constant: **misalignment explains failure; alignment restores coherence.**

Chavan writes not to replace existing disciplines, but to reposition them where intuition and reality finally agree.

Disclaimer and Statement of Originality

Disclaimer

This book is a work of independent research and conceptual interpretation. It is not intended to replace established scientific theories, mathematical formalisms, or professional practice in physics, engineering, medicine, psychology, or any other applied discipline.

The framework presented here—**DELTA**—is offered as a **conceptual and explanatory lens** designed to improve intuition, coherence, and understanding across physical, biological, psychological, and human systems. Wherever classical laws, equations, or models are referenced, they are treated with respect for their predictive success. Any reinterpretation provided is explanatory in nature, not a rejection of empirical validity.

Readers are advised not to substitute the ideas in this book for professional judgment, safety standards, medical advice, or technical protocols. The author assumes no liability for the misuse, misinterpretation, or overextension of the concepts discussed.

This book invites reflection and structural thinking—not belief, compliance, or authority.

Statement of Originality

The ideas presented in this book are the original work of the author.

The **DELTA framework**, including its definitions of delta as misalignment, resolution as the fundamental driver of dynamism, Alignment Threshold Time (ATT), ripple tension, and the reinterpretation of force, friction, pressure, and observation, represents an original synthesis developed through independent study, teaching, and long-term observation.

While the work engages with established domains such as classical mechanics, fluid dynamics, thermodynamics, systems theory, and philosophy, it does so to **reposition causality**, not to replicate or paraphrase existing texts. Any conceptual overlaps with prior theories arise from convergent reasoning, not derivation or duplication.

No part of this book is generated from proprietary sources, unpublished manuscripts, or restricted materials. All interpretations, structures, and frameworks reflect the author's own reasoning and are presented transparently for open examination and critique.

This work does not claim finality. It claims **honest positioning**.

The author welcomes scrutiny, dialogue, and responsible challenge—on the condition that discussion remains grounded in clarity, structure, and reality rather than belief or ideology.

After Resolution

Nothing in this book asks you to believe.

It asks you to notice.

If you look carefully, you may see that events were never driven by hidden hands, invisible forces, or moral preferences. They were shaped by buildup, imbalance, and release.

Outcomes were not rewards. They were resolutions.

There is no privileged observer standing outside reality. You are not separate from what you experience. You are part of the structure that resolves.

When tension builds, something will give. When imbalance accumulates, something will shift. When structure cannot sustain itself, it reorganizes.

This is not punishment. Not destiny. Not randomness.

It is pattern.

Delta is not a doctrine. It is a way of reading that pattern.

If, after closing this book, you find yourself reacting less and observing more... If you begin to ask not "Who caused this?" but "What built this?"... If you see conflict as accumulation and calm as balance rather than victory...

Then the lens has settled.

Nothing mystical has been added. Nothing essential has been removed.

Reality remains what it always was.
Only now, perhaps,
it feels readable.

Further Reading

(From the UED / Chavanian Ecosystem)

The following works expand, contextualize, or apply the structural principles introduced in **Delta**. Each book can be read independently, but together they deepen the reader's ability to perceive reality through resolution rather than force, belief, or abstraction.

Core Structural & Reality Frameworks

- **The Structure of Reality:** A direct companion to *Delta*, examining how outcomes emerge from structure rather than agents, intentions, or hidden forces.
- **Reality Check: Form Is a Guest, Not a Host:** Explores impermanence and appearance, reinforcing Delta's view that form is a consequence—not a controller—of reality.
- **The Cosmos: The Still Universe:** Extends Delta's equilibrium-based worldview to cosmology, challenging motion-centric interpretations of the universe.

Physics Beyond Force-Based Thinking

- **$E \neq mc^2$: The Collapse of Classical Assumptions:** Questions inherited physics assumptions and prepares the reader for non-force, non-object-centered models of reality.
- **Rethinking Energy: Ripple Tension, Not Substance:** Complements Delta's treatment of imbalance and

dissipation by reframing energy as behavior rather than material.

- **The End of Particle Age:** Examines why particle-centric thinking fails at deeper structural levels—an idea implicit throughout *Delta*.

Mind, Intelligence, and Human Perception

- **The Active Mind:** Applies structural dynamics to cognition, showing how resolution governs thought, emotion, and response.
- **The Human Mind: Ripple of Awareness:** Extends Delta's logic into perception and awareness, useful for readers bridging physics and psychology.
- **Chavanian Psychology:** A systemic reinterpretation of psychological functioning through alignment, residue, and consequence.

Unifying Frameworks

- **Universal Energy Dynamics:** Presents the broader framework within which *Delta* operates, for readers seeking a more comprehensive structural map.

Forthcoming & In-Progress Works

(Referenced conceptually but not yet published)

- **Awareness:** A focused exploration of perception as a structural phenomenon rather than a spiritual attribute.
- **Canvas:** Examines context, environment, and boundary conditions as the silent determinants of outcomes.
- **Thermodynamics of Thoughts:** Extends principles of

buildup, dissipation, and equilibrium into cognition and emotional systems.

These forthcoming works will further operationalize the ideas introduced in **Delta**, particularly for readers interested in psychological, behavioral, and societal applications.

Reading Guidance

You do not need to read these books in sequence. Choose based on **domain interest**, not hierarchy.

Delta functions as a lens. The books listed above show what becomes visible once that lens is applied.

Don't miss out!

Visit the website below and you can sign up to receive emails whenever Sandeep Chavan publishes a new book. There's no charge and no obligation.

https://books2read.com/r/B-A-EPGPC-EXACJ

BOOKS 2 READ

Connecting independent readers to independent writers.

Did you love *Delta – The Structure of Change*? Then you should read *The Last Quantum Book*[1] by Sandeep Chavan!

[2]

For over a century, physics has been ruled by an illusion. Quanta, photons, wave–particle duality, collapse of the wavefunction — these phrases became gospel, not because they explained reality, but because they patched equations that otherwise failed. The science worked, the story did not. Generations of students have been trained to recite contradictions: light is both a wave and a particle, matter jumps between shelves, randomness rules at the core of existence. We built technology on these paradoxes, but we never resolved them.

This book argues that we no longer need to. **Quantum was the last great illusion. Ripple Field Dynamics (RFD) is the first true continuity.**

---

1. https://books2read.com/u/m2AYV1

2. https://books2read.com/u/m2AYV1

In *The Last Quantum Book*, physicist and teacher **Sandeep Chavan** dismantles the myths at the foundation of quantum mechanics and introduces a cleaner, sharper framework: ripples in a structured vacuum, governed not by randomness but by thresholds and coherence. Energy is not substance, but consequence. Discreteness is not fundamental, but the footprint of thresholds. Entanglement is not spooky action, but ripple correlation. Collapse is not real, because nothing collapses — alignment simply resolves.

Every major quantum milestone is revisited and reframed in ripple terms:

Blackbody radiation → threshold cutoffs, not packets of energy.**Photoelectric effect** → ripple alignment, not photon absorption.**Lasers** → coherence locking, not photon multiplication.**Semiconductors** → ripple pathways, not electrons in bands.**Superconductivity & tunneling** → large-scale coherence and threshold crossings, not magic pairs or leaking particles.**Quantum computing** → fragile illusions of advantage, dissolving into coherence engineering.

What emerges is not less science, but more — a unified field logic that bridges micro to macro, atom to galaxy, experiment to technology. RFD dissolves paradox without erasing data, offering continuity where physics has long tolerated contradiction.

This is not just a book of physics. It is a call to reimagine science, technology, and civilization itself. Education without paradox. Engineering without packets. Imagination without mystification. A world designed not in fragments, but in ripples.

If the twentieth century was the quantum age, the twenty-first belongs to the ripple.

Quantum was the illusion. Continuity is the truth. This is the last quantum book you will ever need.

Read more at https://www.sandeepjchavan.com/.

www.ingramcontent.com/pod-product-compliance
Lightning Source LLC
LaVergne TN
LVHW090305210926
848480LV00001B/143
* 9 7 9 8 2 3 2 6 8 6 5 3 6 *